BLACKS AND THE MILITARY

Studies in Defense Policy
TITLES IN PRINT

Support Costs in the Defense Budget: The Submerged One-Third
Martin Binkin

The Changing Soviet Navy
Barry M. Blechman

Strategic Forces: Issues for the Mid-Seventies
Alton H. Quanbeck and Barry M. Blechman

U.S. Reserve Forces: The Problem of the Weekend Warrior
Martin Binkin

U.S. Force Structure in NATO: An Alternative
Richard D. Lawrence and Jeffrey Record

U.S. Tactical Air Power: Missions, Forces, and Costs
William D. White

U.S. Nuclear Weapons in Europe: Issues and Alternatives
Jeffrey Record with the assistance of Thomas I. Anderson

The Control of Naval Armaments: Prospects and Possibilities
Barry M. Blechman

Stresses in U.S.-Japanese Security Relations
Fred Greene

The Military Pay Muddle
Martin Binkin

Sizing Up the Soviet Army
Jeffrey Record

Modernizing the Strategic Bomber Force: Why and How
Alton H. Quanbeck and Archie L. Wood with the assistance of Louisa Thoron

Where Does the Marine Corps Go from Here?
Martin Binkin and Jeffrey Record

Deterrence and Defense in Korea: The Role of U.S. Forces
Ralph N. Clough

Women and the Military
Martin Binkin and Shirley J. Bach

The Soviet Military Buildup and U.S. Defense Spending
Barry M. Blechman and others

Soviet Air Power in Transition
Robert P. Berman

Shaping the Defense Civilian Work Force: Economics, Politics, and National Security
Martin Binkin with Herschel Kanter and Rolf H. Clark

The Military Equation in Northeast Asia
Stuart E. Johnson with Joseph A. Yager

Youth or Experience? Manning the Modern Military
Martin Binkin and Irene Kyriakopoulos

Paying the Modern Military
Martin Binkin and Irene Kyriakopoulos

Defense in the 1980s
William W. Kaufmann

The FX Decision: "Another Crucial Moment" in U.S.-China-Taiwan Relations
A. Doak Barnett

Planning Conventional Forces, 1950–80
William W. Kaufmann

Blacks and the Military
Martin Binkin and Mark J. Eitelberg with Alvin J. Schexnider and Marvin M. Smith

STUDIES IN DEFENSE POLICY

BLACKS AND THE MILITARY

Martin Binkin and Mark J. Eitelberg
with Alvin J. Schexnider and Marvin M. Smith

THE BROOKINGS INSTITUTION
Washington, D.C.

Copyright © 1982 by
THE BROOKINGS INSTITUTION
1775 Massachusetts Avenue, N.W., Washington, D.C. 20036

Library of Congress Cataloging in Publication data:

Binkin, Martin, 1928–
   Blacks and the military.

   (Studies in defense policy)
   Includes bibliographical references and index.
     1. United States—Armed Forces—Afro-American soldiers. I. Eitelberg, Mark J.
II. Title. III. Series.
UB418.A47B5   355'.0089'96073   82-70886
ISBN 0-8157-0974-9   AACR2
ISBN 0-8157-0973-0 (pbk.)

9 8 7 6 5 4 3 2 1

Board of Trustees
Robert V. Roosa
Chairman
Andrew Heiskell
Vice Chairman;
Chairman, Executive Committee
Louis W. Cabot
Vice Chairman
Vincent M. Barnett, Jr.
Barton M. Biggs
Frank T. Cary
A. W. Clausen
William T. Coleman, Jr.
Lloyd N. Cutler
Bruce B. Dayton
George M. Elsey
Hanna H. Gray
Huntington Harris
Roger W. Heyns
Bruce K. MacLaury
Robert S. McNamara
Arjay Miller
Herbert P. Patterson
Donald S. Perkins
J. Woodward Redmond
Charles W. Robinson
James D. Robinson III
Henry B. Schacht
Roger D. Semerad
Gerard C. Smith
Phyllis A. Wallace

Honorary Trustees
Eugene R. Black
Robert D. Calkins
Edward W. Carter
Douglas Dillon
John E. Lockwood
William McC. Martin, Jr.
H. Chapman Rose
Robert Brookings Smith
Sydney Stein, Jr.

THE BROOKINGS INSTITUTION is an independent organization devoted to nonpartisan research, education, and publication in economics, government, foreign policy, and the social sciences generally. Its principal purposes are to aid in the development of sound public policies and to promote public understanding of issues of national importance.

The Institution was founded on December 8, 1927, to merge the activities of the Institute for Government Research, founded in 1916, the Institute of Economics, founded in 1922, and the Robert Brookings Graduate School of Economics and Government, founded in 1924.

The Board of Trustees is responsible for the general administration of the Institution, while the immediate direction of the policies, program, and staff is vested in the President, assisted by an advisory committee of the officers and staff. The by-laws of the Institution state: "It is the function of the Trustees to make possible the conduct of scientific research, and publication, under the most favorable conditions, and to safeguard the independence of the research staff in the pursuit of their studies and in the publication of the results of such studies. It is not a part of their function to determine, control, or influence the conduct of particular investigations or the conclusions reached."

The President bears final responsibility for the decision to publish a manuscript as a Brookings book. In reaching his judgment on the competence, accuracy, and objectivity of each study, the President is advised by the director of the appropriate research program and weighs the views of a panel of expert outside readers who report to him in confidence on the quality of the work. Publication of a work signifies that it is deemed a competent treatment worthy of public consideration but does not imply endorsement of conclusions or recommendations.

The Institution maintains its position of neutrality on issues of public policy in order to safeguard the intellectual freedom of the staff. Hence interpretations or conclusions in Brookings publications should be understood to be solely those of the authors and should not be attributed to the Institution, to its trustees, officers, or other staff members, or to the organizations that support its research.

FOREWORD

RACE RELATIONS have been among the most emotional, complex, and enduring sociopolitical issues facing the United States and, indeed, the world community. Thus it comes as no surprise that the nation's military establishment, even three decades after pioneering racial integration, is not free of controversy.

But unlike questions of discrimination that dominated earlier debates, the contemporary wrangle centers on issues related to the dramatic growth in the 1970s in the participation by blacks in the military services. In contrast to the racial proportionality that characterized U.S. armed forces throughout most of the nation's history, the 410,000 blacks under arms in 1981 represent about 20 percent of all military personnel, a proportion far greater than the 11 or 12 percent of the total population that is black. Moreover, blacks make up a still larger share of the ground combat forces: one of every three Army GIs is black, as is one of every five enlisted marines. The public reaction to these developments has been mixed.

Many Americans look with approval on the growth of black participation in military service, since it often affords young blacks educational, social, and financial opportunities that constitute a bridge to a better life not otherwise available to them. The fact that two of every five eligible young black males now enter the military attests to its importance to them as an employer.

But for other Americans, these opportunities are outweighed by the disproportionate imposition of the burden of defense on a segment of the population that has not enjoyed a fair share of society's benefits. From this perspective, the likelihood that blacks would suffer a third—and perhaps half—of the combat fatalities in the initial stages of conflict is considered immoral, unethical, or otherwise contrary to the precepts of democratic institutions.

Some also worry that military forces with such a high fraction of blacks entail risks to U.S. national security. A socially unrepresentative force, it is argued, may lack the cohesion considered vital to combat effectiveness. Others have suggested that such a force might be unreliable if it were deployed in situations that tested the allegiance of its minority members. And some have even expressed concern that a large proportion of blacks may raise questions about the status of U.S. fighting forces, as judged by the American public, the nation's allies, and its adversaries.

These controversial questions are addressed in this study, the twenty-sixth in the Brookings Studies in Defense Policy series. The authors first trace the history of black participation in the U.S. armed forces and describe the changes in the racial composition that marked the 1970s. A discussion of the social dilemma—the burdens versus benefits of military service—is followed by an examination of the effectiveness, reliability, and public perceptions of racially unbalanced military forces. A separate chapter is devoted to an analysis of the implications for the military's racial mix of demographic, economic, and technological trends and the possible effects of a return to some form of conscription.

The authors have not attempted to settle the questions. Their purposes are less ambitious: to draw the issues, to present the available evidence, to identify gaps in our knowledge, to stimulate further research, and to promote an informed debate. In the process, they have analyzed and presented a wealth of data, much of which has not previously appeared in the public record.

Martin Binkin, a senior fellow in the Brookings Foreign Policy Studies program, is author or coauthor of eight previous Studies in Defense Policy. Mark J. Eitelberg is a senior scientist with the Human Resources Research Organization. Alvin J. Schexnider is associate dean of the School of Community and Public Affairs at Virginia Commonwealth University, and Marvin M. Smith is a research associate in the Brookings Economic Studies program.

The book has benefited greatly from the comments of Buddy Beck, Barry M. Blechman, William K. Brehm, Major General Frederic E. Davison (U.S. Army, retired), Thomas W. Fuller, Eli Ginzberg, Richard O. Hope, Daniel Huck, Charles C. Moskos, General Bruce Palmer, Jr. (U.S. Army, retired), David R. Segal, Wayne S. Sellman, Raphael Thelwell, Brian K. Waters, Togo D. West, Jr., and Adam Yarmolinsky. The authors are indebted to Kenneth C. Scheflen, Zahava D. Doering,

Alex Sinaiko, and Leslie W. Willis of the Defense Manpower Data Center for their generous support.

The authors are also grateful to John D. Steinbruner, director of the Brookings Foreign Policy Studies program, who provided valuable guidance; to Elizabeth H. Cross, who edited the manuscript; to Clifford A. Wright, who verified its factual content; to Florence Robinson, who prepared the index; and to Ann M. Ziegler, who patiently and skillfully processed the manuscript.

The Institution gratefully acknowledges the assistance of the Ford Foundation and the Edna McConnell Clark Foundation, whose grants supported this study. The views expressed herein are those of the authors and should not be ascribed to the persons who provided data or who commented on the manuscript, to the Ford or Edna McConnell Clark Foundations, or to the trustees, officers, or other staff members of the Brookings Institution.

BRUCE K. MACLAURY
President

April 1982
Washington, D.C.

CONTENTS

1. **The Setting** — 1
2. **From Bunker Hill to Vietnam** — 11
   The First Three Centuries *11*
   World War II and Korea: The Fight for the Right to Fight *18*
   Vietnam and the New Era of Racial Representation *32*
3. **Blacks in the Post-Vietnam Army** — 39
   Recruitment and Retention *39*
   Profile of Black Volunteers *45*
4. **Benefits versus Burdens** — 62
   Benefits of Military Service *65*
   Burdens of Military Service *75*
5. **Racial Composition and National Security** — 84
   Individual Capabilities *84*
   Group Performance *98*
   Foreign Perceptions and Reactions *115*
6. **Looking Ahead** — 120
   Ethnodemographic Trends *120*
   Tested Abilities and Aptitudes *121*
   Economic Factors *125*
   Technological Trends *132*
   The Prospective Military Buildup *135*
   Military Manpower Policies *136*
7. **The Policy Dilemma** — 152

   Appendixes
   A. Racial-Ethnic Categories in the Armed Forces *160*
   B. Statistical Tables *163*

   Index — 185

Tables

3-1. Blacks as a Percentage of the Armed Forces, by Service, Selected Fiscal Years, 1942–81 — 42

3-2. Black Enlisted Entrants to the Army, by Sex, Selected Fiscal Years, 1954–81 — 45

3-3. Percentage Distribution of Male Enlisted Entrants to All Services, by AFQT Category and Race, Selected Years, 1953–81 — 48

3-4. Selected Characteristics of Males in the Armed Forces and Employed Male Civilians Aged 18 to 21, by Race, 1979 — 50

3-5. Trends in Disciplinary Incidents in the Armed Forces, by Race and Sex, Fiscal Years 1978–81 — 52

3-6. Crime Rates of Army Personnel, by Race, Fiscal Years 1978–80 — 53

3-7. Blacks as a Percentage of Male Enlisted Personnel Assigned to Major Occupational Areas in the Army and All Services, Selected Years, 1964–81 — 56

3-8. Blacks as a Percentage of Total Enlisted Personnel, by Pay Grade and Service, September 1981 — 59

3-9. Blacks as a Percentage of Officers Assigned to Major Occupational Areas, by Service and Sex, September 1981 — 60

3-10. Blacks as a Percentage of Total Officers, by Pay Grade and Service, September 1981 — 61

4-1. Military Participation Rates of Male Youths Born between 1957 and 1962, by Race and Education — 66

4-2. Unemployment Rates, by Race and Age, Selected Years, 1955–81 — 68

4-3. Employment–Population Ratios, by Race and Age, Selected Years, 1955–81 — 69

4-4. Distribution of Male Enlisted Personnel, All Services, by Major Occupational Category and Race, September 1981 — 74

4-5. Army Combat Deaths in Vietnam, by Race, 1961–72 — 76

4-6. Blacks as a Percentage of Total Enlisted Personnel Assigned to Army Divisions and Selected Brigades and Regiments, September 1980 — 79

5-1. Males Aged 18 to 21 Who Completed High School, by Race, Selected Years, 1940–80 — 94

5-2. Estimated Percentage Distribution of Males' Entry Test Scores Referenced to 1944 Standard, by Race, Selected Years, 1941–80 — 95

5-3. Aptitude Subtest Components, by Army Occupational Area — 95

5-4. Comparative Means and Standard Deviations of Raw Scores on Armed Services Vocational Aptitude Battery, 18- to 23-Year-Old Males, by Race, 1980 — 97

5-5. Comparative Means and Standard Deviations of Percentile Scores of 18- to 23-Year-Olds on Selected Aptitude Composites, by Race and Sex, 1980 — 97

5-6. Males Aged 18 to 23 Eligible for Enlistment Based on Fiscal 1981 Aptitude Test Standards, by Level of Education and Race — 98

6-1. Projected U.S. Population Aged 18 to 21, by Sex and Race, Selected Years, 1981–95 — 121

6-2. Comparative Means and Standard Deviations of American College Testing Program Assessment Scores of College-bound Students, Selected School Years, 1970–80 ... 123
6-3. Mean Percentages and Changes in Correct Responses for Seventeen-Year-Olds in Reading Performance, by Race, 1971 and 1980 ... 124
6-4. Comparison of Rates of Unemployment Projected by the Reagan Administration and the Congressional Budget Office, 1982–87 ... 127
6-5. Distribution of Trained Military Enlisted Personnel by Major Occupational Category, All Services, 1945, 1957, and 1981 ... 133
6-6. Army Recruits Who Enlisted under Bonus Program, by Race, 1976, 1978, and 1980 ... 140
6-7. Percentage Distribution of Male Army Entrants, Fiscal Year 1981, and Two Conscription Alternatives, by Race, AFQT Category, and Level of Education ... 146

Appendix Tables

A-1. Military Personnel by Race or Ethnic Category, September 1981 ... 162
A-2. Percentage Distribution by Race or Ethnic Category, Military Personnel, September 1981, and Total Population, 1980 ... 162
B-1. Blacks as a Percentage of Selected Reserve Forces, by Component, Fiscal Years 1972–81 ... 164
B-2. Army Reenlistment Rates, by Race and Career Status, and Racial Composition of All Army Reenlistments, Fiscal Years 1972–81 ... 164
B-3. Percentage of New Recruits with a High School Diploma, by Race and Service, 1972–81 ... 165
B-4. Estimated Family Income, by Race, from 1979 Defense Department Survey of Personnel Entering Military Service ... 166
B-5. Military Personnel Not Completing Their First Enlistment Period, by Education, Race, and Sex, Fiscal Years 1972–78 ... 167
B-6. Trends in Disciplinary Incidents, by Service, Race, and Sex, Fiscal Years 1979–81 ... 168
B-7. Recruits Who Received Moral Waivers, by Type of Waiver, Service, and Race, Fiscal Year 1981 ... 169
B-8. Distribution of Enlisted Personnel Discharged from the Armed Forces, by Character of Service, Race, and Branch of Service, Fiscal Year 1980 ... 170
B-9. Enlisted Separation Rates, by Cause of Separation, Race, and Sex, Fiscal Year 1980 ... 171
B-10. Blacks as a Percentage of Male Enlisted Personnel Assigned to Occupational Areas, by Service, Selected Years, 1964–81 ... 172
B-11. Blacks as a Percentage of Army Male Enlisted Personnel Assigned to the Twenty Most Common Occupational Subgroups, Fiscal Years 1972, 1976, and 1981 ... 173
B-12. Black Male Enlisted Personnel in the Army, by Selected Primary Military Occupational Specialty and Occupational Area, September 1981 ... 174
B-13. Sources of Commissions for Officers on Active Duty, September 1981 ... 177

B-14. Reasons Given by Recruits for Joining the Military, by Race, 1979   178
B-15. Percentage of Black and Other Minority Personnel in Selected Army Units, December 1980   179
B-16. Percentage of Black and Other Minority Enlisted Personnel in Marine Corps Units, December 1980   182

Figure

4-1. Comparison of Mean Annual Earnings of Male Civilians, by Race and Age Group, and of Military Enlisted Personnel, by Age Group, 1979   72

CHAPTER ONE

THE SETTING

IN 1968 General Lewis Hershey, director of the Selective Service System, asserted that "the System is representative of the American people, as clear an example as exists today of government of the people, by the people, and for the people. . . . The system as constituted invades all economic levels, all educational institutions, all geographic areas and all ethnic groups."[1] Hershey was responding to critics who found an inequitable distribution of the "burden of defense" among the nation's youth. However, Pentagon statistics were showing that black Americans were more likely to be drafted, to be sent to Vietnam, to serve in high-risk combat units, and consequently to be killed or wounded.[2] Also, an array of deferments and disqualifications—for getting married, having a child, enrolling in college, teaching in public school, joining the Peace Corps, or failing the induction physical examination—left numerous ways for young men to avoid the draft, and those who did, it appeared, were mainly the white, better-educated children of comfortable families.

By 1969 the end of conscription seemed inevitable. The rising tide of public opinion favored President Nixon's 1968 campaign promise to "prepare for the day when the draft can be phased out of American life."[3] At the same time, prevailing views of the relationship between

---

1. Lewis B. Hershey, "The Operation of the Selective Service System," *Current History*, vol. 55 (July 1968), p. 50.
2. "How Negro Americans Perform in Vietnam," *U.S. News & World Report*, August 15, 1966, pp. 60–64. However, see Gilbert Badillo and G. David Curry, "The Social Incidence of Vietnam Casualties: Social Class or Race," *Armed Forces and Society*, vol. 2 (Spring 1976), pp. 397–406.
3. Cited in Melvin R. Laird, *Report to the President: Progress in Ending the Draft and Achieving the All-Volunteer Force* (Department of Defense, Office of the Secretary of Defense, July 20, 1972), p. 1. See also Richard M. Nixon, "The All-Volunteer Armed Force," address given over the CBS radio network, October 17, 1968, in Gerald Leinwand, ed., *The Draft* (Pocket Books, 1979), pp. 96–108.

the military and society were undergoing significant changes. First, the war in Vietnam (along with increased draft calls) gave the armed forces a new and higher level of visibility. The seemingly endless war, the daily body counts and reports of missing persons, selective service reform, and the movement to end conscription were important public concerns, and concurrently, "quota consciousness" was becoming a major social and political issue. The civil rights movement, women's liberation, the welfare rights movement, Supreme Court decisions, the War on Poverty, and federal legislation to create a "balanced society" contributed to a heightened awareness of group participation and "statistical parity" in all sectors of society.

In 1970 the stage was set for serious debate concerning the practicality of an all-volunteer force—not only whether it was feasible, but whether a volunteer system could amend the social injustices of a less than equitable draft. The equity issue became a primary argument of critics of voluntary recruitment, who claimed that abolition of the draft would further insulate the better-educated sons of middle- and upper-class families from military service and the horrors of war.[4]

The first negative reactions to the introduction of the plan for "zero-draft" calls, however, generally had to do with national security and the means of maintaining a mass armed force—the major reasons given for instituting conscription. There were some references to the issues of proportional "representation" in early discussions, but it was the final report of the President's Commission on an All-Volunteer Armed Force (often referred to as the Gates Commission after its chairman, former Secretary of Defense Thomas S. Gates, Jr.) and its treatment of objections to the all-volunteer force that provided the first official recognition of possible representation problems.

The Gates Commission report identified and then dismissed several contemporary issues that were directly related to questions of complete citizen participation: (1) an all-volunteer force will "undermine patriotism by weakening the traditional belief that each citizen has a moral

---

4. This particular comment is attributed to Senator Edward M. Kennedy. See, for example, statement by Kennedy before the Senate Armed Services Committee cited in *The Power of the Pentagon* (Washington, D.C.: Congressional Quarterly, 1972), p. 50. See also James W. Davis, Jr., and Kenneth M. Dolbeare, *Little Groups of Neighbors: The Selective Service System* (Markham, 1968); Harry A. Marmion, *The Case Against a Volunteer Army* (Quadrangle Books, 1971); Blair Clark, "The Question Is What Kind of Army?" *Harper's*, September 1969, pp. 80–83; and "The Question of an All-Volunteer U.S. Armed Force: Pro & Con," *Congressional Digest*, vol. 50 (May 1971).

responsibility to serve his country";[5] (2) the presence of self-selected, "undesirable psychological types, men inclined to use force and violence to solve problems," will isolate the military from society and threaten "civilian authority, our freedom, and our democratic institutions";[6] (3) the volunteer force will be all black or dominated by servicemen from low-income backgrounds, "motivated primarily by monetary rewards rather than patriotism";[7] (4) the volunteer force will lead to a decline in patriotism, a decline in popular concern about foreign policy, and an increase in the likelihood of military adventurism;[8] and (5) there will be a general erosion of military effectiveness "because not enough highly qualified youths will be likely to enlist and pursue military careers," further causing an erosion "of public support of armed services" and a decline in "the prestige and dignity of the services."[9]

During the transition from draft to volunteer force, the major concern for most policymakers was "quantity and quality." Issues of representation were secondary since, to be effective, the armed forces would first have to attract adequate *numbers* of qualified volunteers. However, in 1972 Defense Secretary Melvin Laird did point out that "long range ... we do not foresee any significant difference between the racial composition of the All-Volunteer Force and the racial composition of the Nation"; and charges that it will be dominated by mercenaries or be all black or be dominated by low-income youth are "false and unfounded claims."[10] Indeed, Laird reported, "we are determined that the All-Volunteer Force shall have broad appeal to young men and women in all racial, ethnic, and economic backgrounds."[11]

When it became apparent that quantitative requirements could be achieved under volunteer conditions, attention shifted to qualitative considerations and the finer points of military representation. By the end of 1974 it was obvious that certain social groups were not enlisting at predicted levels; the "broad appeal" of military service did not extend quite as far as many defense analysts and devotees of voluntarism had envisioned. The most conspicuous statistic was the rapid surge in the

5. *The Report of the President's Commission on an All-Volunteer Armed Force* (Macmillan, 1970), p. 13.
6. Ibid., pp. 131, 14.
7. Ibid., p. 16.
8. Ibid., pp. 16–17.
9. Ibid., pp. 18, 136.
10. Laird, *Report to the President*, pp. 26, 8.
11. Ibid., p. 26.

proportion of black Army recruits to an unprecedented high of 27 percent—a substantial increase from the 15 percent level experienced during the last year of the draft, and more than double the percentage of black recruits of just a few years before, when the Gates Commission recommended the formation of an all-volunteer military and dismissed the likelihood of a racially unbalanced force. In all services combined the proportion of black rank and file stood at about 16 percent in 1974, but total black enlistments had risen from 13 percent in 1970 to 21 percent only four years later—and a combination of factors suggested even higher proportions in the years ahead.[12]

"We are watching these figures," wrote Assistant Secretary of Defense William Brehm, "but are not now concerned about them for one important reason: the Department of Defense sets high entrance standards for enlistment—standards designed to assure that an applicant can perform a military mission as a member of a team."[13] But the individual services did not completely share that view. In March 1975 Army Secretary Howard H. Callaway described the Army's manpower recruitment goals, taking the issue of "representation" to its idealistic extreme:

What we seek, and need, are quality soldiers—men and women—who are representative of the overall population. Ideally, we would like to have at least one person from every block in every city, one from every rural delivery route, and one from every street in every small town. Our obligation to the American people is to strive to field an Army which is both representative of them and acceptable to them.[14]

The nation needs, Callaway explained, "an army broadly representative of all Americans which, to the extent possible, would contain roughly the same representative percentages of people of all ethnic groups, and the same percentages at various income levels and educational levels."[15] The Army's top personnel officer said, "We believe

---

12. Kenneth J. Coffey and others, "The Impact of Socio-Economic Composition in the All Volunteer Force," in *Defense Manpower Commission Staff Studies and Supporting Papers,* vol. 3: *Military Recruitment and Accessions and the Future of the All Volunteer Force* (Government Printing Office, 1976), p. E-12.

13. William K. Brehm, "A Special Status Report: All-Volunteer Force," *Commanders Digest,* February 28, 1974. The Department of Defense took care *not* to establish a position on racial balance on the ground that the social composition of the armed forces was considered the public's business, not the Pentagon's.

14. In *Department of Defense Appropriations, Fiscal Year 1976,* Hearings before the Senate Committee on Appropriations, 94 Cong. 1 sess. (GPO, 1975), pt. 2, p. 13.

15. Ibid., p. 105.

that these quality personnel should be representative of all regional, economic, and racial segments of society";[16] an Army that is "generally representative of the American people . . . in the racial, geographic, and socio-economic sense," echoed Assistant Secretary of the Army for Manpower and Reserve Affairs Donald G. Brotzman.[17] Toward that end, in 1975 "the Army redistributed its recruiting force with a stated objective of achieving better geographical representation among recruits."[18] This move, called the "Callaway shift" by insiders, transferred some recruiters out of heavily black areas, "although it would have been more efficient and cost effective to concentrate recruiters in certain 'pro-Army' areas of the country."[19]

The Navy was also accused in 1975 of implementing "policies which directly limit the enlistment of blacks."[20] For example, the Navy's quota system allowed recruiters to sign up only one "category IV" (the lowest acceptable level) volunteer for every ten whose tests indicated they could do well in technical school. Although the 10-to-1 ratio applied to whites and blacks alike, the system was actually loaded against blacks since proportionately more black applicants normally scored in the lower aptitude categories and proportionately fewer could therefore be accepted by the Navy. During the same year the Marine Corps was challenged for giving recruiters secret racial quotas, and service entrance and placement tests were denounced as racially biased.[21]

16. Testimony of Lieutenant General Harold G. Moore in ibid., p. 619.
17. Quoted in Kenneth J. Coffey and Frederick J. Reeg, "Representational Policy in the U.S. Armed Forces," in *Defense Manpower Commission Staff Studies and Supporting Papers,* vol. 3, p. D-13. See also "Statements of Assistant Secretary of Defense William K. Brehm before Subcommittee on Manpower and Personnel of the Senate Armed Services Committee" (Office of the Assistant Secretary of Defense for Manpower and Reserve Affairs, February 6, 1976), p. 43.
18. Coffey and Reeg, "Representational Policy," p. D-16.
19. Ibid., p. D-17.
20. George C. Wilson, "Bias in Recruiting Laid to 4 Services," *Washington Post,* June 8, 1976.
21. Ibid. Again, in 1979, the Navy was accused of practicing "blatant" and "illegal" racial discrimination in its entrance standards for volunteers. Several congressmen along with the American Civil Liberties Union based their accusations this time on the Navy's requirement that at least 75 percent of the males in any racial category who are accepted into the service must either have a high school diploma or achieve a certain score on the aptitude tests. (For example, 75 blacks would have to achieve acceptably high scores on the aptitude tests before 25 blacks with low scores could be accepted, and the same for other races.) However, for a variety of reasons blacks and other minorities generally score lower on the aptitude tests. "Blacks and other minorities are being skillfully steered away from the military," Congressman Ronald V. Dellums of California charged. This is "very

The Army has always been the focus of discussions of military representation since it requires the greatest manpower, is generally considered the least glamorous branch of the armed forces, and is consequently the least socially "representative" service.[22]

When Army Secretary Clifford L. Alexander (the first black appointed to the position) took office in 1977, he answered critics of the growing racial imbalance by contending that the number of blacks in the Army is "immaterial": "Who is going to play God and set a quota?" Alexander has continued to maintain that the problem lies "outside the services"; you have to ask "why there is almost 40 percent unemployment among black teenagers before you ask why they enlist or why they re-up."[23] Although Alexander believed the Army of the late 1970s was "the best ever assembled,"[24] he noted that "minority and female representation" in certain occupations and on "high level staffs" could be improved. "We can do better," he wrote.[25] Meanwhile, the proportion of blacks in the armed forces continued to grow, nudging 20 percent of total enlisted personnel and reaching all-time highs of 33 percent and 22 percent in the Army and Marine Corps enlisted ranks, respectively, by the end of the decade.

It did not take long for Secretary of the Army John D. Marsh, Jr., to get involved in the issue. During his confirmation hearings in 1981, Marsh testified:

I happen to feel that service in the U.S. Army is not only a privilege, it is a duty of every citizen. . . . I also believe that it is not fair for that burden to be

---

definitely a quota system," Congressman Don Edwards of California added, which "discriminates against minorities" and is "unconstitutional" and "illegal." George C. Wilson, "Navy Is Accused of Bias in Entrance Standards," *Washington Post,* June 14, 1979.

22. As of September 1981 approximately 38 percent of all active duty military personnel were in the Army. During the peak manpower period of the Vietnam War (1968), Army personnel constituted over 44 percent of the total active duty military and about 45 percent of the total active duty *enlisted* force. Department of Defense, *Selected Manpower Statistics* (Directorate for Information, Operations, and Reports, 1978), pp. 20, 26.

23. David Binder, "Army Head Favors Volunteers," *New York Times,* February 11, 1977. See also George C. Wilson, "Blacks in Army Increase 50 Percent Since Draft," *Washington Post,* May 2, 1978.

24. Interview on "America's Black Forum," Station WMAL-TV, Washington, D.C., April 10, 1977.

25. *Equal Opportunity: Second Annual Assessment of Programs* (Office of the Deputy Chief of Staff for Personnel, Department of the Army, 1978) (letter accompanying report, dated April 1978).

unequally borne in our society. . . . I do think that a national military force should represent as much as it might some cross-section of our country.[26]

But responding to a question in a postconfirmation interview concerning the point at which the growing proportion of blacks in the Army would become a factor, Marsh replied: "I don't believe in quotas. . . . It's my own view that we work really with what we have and I don't think that I should try to hypothesize problems that don't exist. . . . I don't see any problem in our ratios at the present time."[27]

While the changing racial mix may have been ignored in formal government channels, it did not escape the attention of outside commentators. Some scholars contend that a military force that fails to represent society poses a threat not only to civilian control of the military but to its effectiveness as well. Some national leaders—both black and white—hold that a disproportionately black force puts an unfair burden on black Americans, particularly in the initial stages of military hostilities. Other observers question the reliability of an unrepresentative force, particularly when such a force might be assigned to missions (domestic or foreign) in which their representativeness would create an issue. And some have even suggested that an increasingly black force has adversely influenced the caliber of white recruits.

Whatever the validity of these viewpoints, constituencies have formed around them, attributable in part to the influence of the popular media. The *New York Times,* for example, repeatedly noted the "drift toward a heavily black Army" in its criticisms of the all-volunteer military during the late 1970s. As early as 1975, the *Times* warned:

In a population 11 percent black, the proportion of blacks in the Army as a whole has risen by almost half since 1971 to a current level of 20 percent, and even these figures understate the real problem. . . . The end result can be a ground force so largely made up of blacks as to destroy the integration goal.[28]

In May 1978 the *Times* again singled out the representation problems of the military:

It is now an Army with substandard education, heavy racial imbalance and a drop-out rate double that of the draft era. . . . Eliminating the Selective Service System has not in fact eliminated the inequities that helped spur agitation against the draft during the Vietnam War. . . . There are more poor in the Army now,

26. *Nomination of John O. Marsh, Jr., to be Secretary of the Army,* Hearing before the Senate Armed Services Committee, 97 Cong. 1 sess. (GPO, 1981), p. 13.
27. "Marsh Wants More Help for Reserves," *Army Times,* April 6, 1981.
28. *New York Times,* February 5, 1975.

not less. The percentage of blacks among Army enlisted men in 1971 was 13 percent, about the same as in the nation; it is now double that among Army recruits. Among officers, the proportion of blacks is only 6.3 percent.[29]

And in 1979: "The strength, quality and cost of the volunteer force are all sources of worry," but the "more worrisome" problem is the fact that the "Army is no longer even roughly a cross section of the Nation." Volunteers "are coming far more heavily from the ranks of the poor, the unemployed and the undereducated than did even the troops in Vietnam."[30]

"The services are growing dramatically unrepresentative of the nation," *Time* magazine found. "A number of military experts argue that while it is true that peacetime service offers to minorities opportunities for educational and social advancement, these advantages fade quickly during a war." And "the high number of blacks in uniform would inevitably result . . . in a disproportionate number of black fatalities."[31]

"The disproportionate number of poor, uneducated and blacks" is a "condition that exposes the nation to the charge of turning over its defense to the most disadvantaged elements of society while relieving the middle and upper classes from participation in the dangerous and highly unpleasant business of fighting our wars," a former chairman of the Joint Chiefs of Staff argued in the *Washington Post*.[32] The "ambitious experiment" to maintain a military force composed entirely of volunteers "has not worked well," an editorial in *Time* magazine concluded. "The racial balance does not reflect that of the nation." The draft should therefore be restored, stated the editorial, since it would provide the Army with "a more representative cross section" of American youth.[33]

In early 1979 a *New York Times* reporter observed that "many critics, both liberals and conservatives alike, believe that the military has become

29. "Can We Afford a Volunteer Army?" editorial in the *New York Times*, May 18, 1978.
30. "Misgivings About the Volunteer Army," editorial in the *New York Times*, January 2, 1979.
31. "Who'll Fight for America? (The Manpower Crisis)," *Time*, June 9, 1980, p. 25.
32. General Maxwell D. Taylor, "Is the Army Fit to Fight?" *Washington Post*, May 12, 1981. See also two replies to Taylor: Clifford L. Alexander, "Now Is Not the Time to Draft," *Washington Post*, May 14, 1981; and Lawrence J. Korb, "Volunteer Army: It Deserves a Fair Chance," *Washington Post*, June 9, 1981.
33. "Needed: Money, Ships, Pilots—and the Draft," *Time*, February 23, 1981, p. 56. And a *Washington Post* columnist observed: "Defending the United States is just as much the responsibility of Nick and Adam as it is of Jose and Tyrone." Mark Shields, "Checkbook Patriotism Won't Do," *Washington Post*, March 6, 1981.

totally unrepresentative of American society. . . . As they do periodically, these criticisms have led to discussion of reviving the draft."[34] Another commentator put it more bluntly: "Uncle Sam does want you—if you're white, bright, and ready to fight. And that may be why he's thinking about putting the draft back to work: The U.S. Army is short on white men with managerial or technical know-how."[35]

Advocates of the all-volunteer structure frequently find themselves on the defensive—fending off the charges of detractors and fighting to save a concept that can perhaps function effectively under the proper conditions. President Ronald Reagan thus sees "a new spirit abroad in our land" now bringing to the military "a decided rise in quality as measured by educational and testing attainment."[36] And the Defense Department continues to maintain that "while not without problems, the AVF [all-volunteer force] is working."[37] Still, the popular media brood over an armed force full of losers and social outcasts, disadvantaged minorities, and "hired guns" conscripted through economic poverty to bear arms by an employer of last resort. This perception has helped to push the all-volunteer force closer than ever before to a new form of conscription. "Some critics . . . complain that an all-volunteer military will become increasingly unrepresentative of American society," *Newsweek* notes. "If the President's plans for a massive defense

---

34. Bernard Weinraub, " 'National Service'—An Old Idea Gets New Life," *New York Times*, February 4, 1979. Because the major shift in the racial mix happened to occur under a volunteer recruitment system, it is frequently but inappropriately cited as proof of the failure of the concept. But if the proponents of voluntary service had not been so *emphatic* in their predictions of "proportional representation," perhaps the reactions of critics and skeptics would not have been so severe. The Gates Commission had left little room for doubt; their "best projections for the future" were that blacks would constitute 14.9 percent of all enlisted males and that the proportion of black enlistees in the Army would be approximately 18.8 percent by 1980. "To be sure, these are estimates," the commission asserted, "but even extreme assumptions would not change the figures drastically." See *Report of the President's Commission on an All-Volunteer Armed Force*, pp. 15, 147. For the argument that "the increasing number of blacks in the enlisted accession of the 1970s would probably have taken place *even in the presence of the draft*," see Richard V. L. Cooper, *Military Manpower and the All-Volunteer Force*, R-1450-ARPA (Santa Monica: Rand Corp., 1977), p. 219; emphasis in the original.

35. Joseph Kelley, "Behind the Push to Revive the Draft," *The Progressive*, May 1980, reprinted in Jason Berger, ed., *The Military Draft* (H. W. Wilson, 1981). Quotation appears on pp. 18–19.

36. "Text of President's West Point Speech," *Army Times*, June 8, 1981, p. 53.

37. Department of Defense, "Fiscal Year 1981 Results" (Office of the Assistant Secretary of Defense for Manpower, Reserve Affairs, and Logistics, November 1981).

buildup move ahead on schedule, a return to the draft seems all but unevitable."[38]

New military pay raises, intensified recruiting efforts, a surge of national pride in the wake of foreign events, a depressed civilian job market (especially for teenagers), and other factors have combined to make the early 1980s something of a recruiting success for the all-volunteer force. Some faultfinders have softened their blows as "the downward spiral of quality"[39] appears to have gone into reverse. Yet criticisms of the quality of soldiers and racial imbalances still serve as the broadsword of those who would prefer to see a revival of the draft.

Is the concern justified? Is it appropriate for the nation's disadvantaged minorities to bear the burden of protecting its security? Is an armed force that fails to represent society less effective, less reliable, or less legitimate? How is the racial composition of the armed forces likely to be affected by the demographics and economics of the 1980s, by changes in military pay and benefits, or by a return to some form of conscription?

Caught in a crossfire of emotions, these questions have so far escaped objective scrutiny and informed debate. This study is intended to promote a better public understanding of the issues. It does not attempt to judge whether the current racial composition of the U.S. armed forces or the initiatives already undertaken that may change it are appropriate. The study's purposes are more modest: to identify the range of concerns, to examine the evidence on both sides of the questions, and to stimulate further research and debate.

Although many of the questions apply to some extent to all minority groups, the focus is on blacks, first, because blacks constitute by far the largest of the racial or ethnic minority groups in the armed forces (see appendix A); second, because black-white relations have long been one of the major sociopolitical issues facing the nation; and third, because data on other racial or ethnic groups are limited. Where possible, the analyses in this study are extended to other minority groups, particularly Hispanics.

38. "Why a Draft Seems Certain," *Newsweek,* June 8, 1981, p. 39. See also Marvin Stone, "Is a Draft Inevitable?" *U.S. News & World Report,* July 13, 1981, p. 80.
39. "Today's American Army," *The Economist,* April 25, 1981, p. 24.

CHAPTER TWO

FROM BUNKER HILL TO VIETNAM

TODAY the U.S. military establishment stands in sharp and favorable contrast with other institutions in American society in making headway against racism. Racial integration in the armed forces—admittedly a difficult undertaking and one that is far from complete—was a precursor of the civil rights movement. Yet for most of its history the nation's military has been a white bastion whose personnel policies mirrored the racial prejudices prevalent in the rest of society. Indeed, the history of class privilege and racial injustice is faithfully reflected in the history of the American military.

The nation, at various times and under various circumstances, has denied members of certain social categories entrance into military service when it was important to them to serve and has protected members of other groups when it was important to them not to serve. The black experience in the American armed forces has likewise been marked by policies of exclusion during periods of peace and expedient acceptance during mobilization for war. Although blacks have taken part in all the nation's wars, the armed forces openly sustained the indignities and humiliation, the discrimination, and the stereotypes of racial inferiority until the middle of the twentieth century.

The First Three Centuries

In the original militias of colonial America every available man—white or black, freedman or slave—was to help defend the domestic order against Indian uprisings, European transgressors, and other threats to peace. However, colonial leaders soon realized that domestic order was constantly threatened by the possibility of slave revolts. Persuaded by the fear that free black militiamen might support such insurrections and a related apprehension about training slaves in the use of arms, the American colonies developed a policy of excluding blacks from military service.[1]

The first such provision was instituted in 1639 by the colony of Virginia. In 1656, Massachusetts passed a similar measure, and in 1661, four years after an uprising of blacks and Indians in Hartford, Connecticut followed suit. The other colonies later restricted the participation of blacks in military affairs "lest our slaves when armed might become our masters."[2]

Subsequent regulations stipulated that free blacks could serve only as drummers, fifers, laborers, and in other ancillary positions that did not require them to bear arms. Nevertheless, military necessity frequently compelled the colonies to overlook or rescind provisions excluding blacks from the militia, and some colonies even promised freedom to slaves who performed well in battle. The prospect of freedom thus attracted slaves to serve in the colonial forces, and free blacks who distinguished themselves hoped this would lift them from their low social status.[3]

The Revolutionary War

Black militiamen participated in the early battles of the American Revolution, including Lexington, Concord, Ticonderoga, and Bunker Hill. Yet fears of a possible insurrection by blacks again surfaced. Powerful slaveholders also objected to militia recruitment policies that offered runaways a refuge and other slaves a pathway to eventual freedom. White supremacists viewed blacks as inherently inferior and untrustworthy, and some colonial leaders considered it morally wrong to ask slaves and former slaves to share the burdens of defense. In response to these pressures and with the support of the Continental Congress, General Washington issued an order in 1775 prohibiting any new enlistments of blacks (but allowing blacks who were already in the army to remain there).

    1. The discussion of black participation in the militias of colonial America through the Civil War draws heavily on the following sources: Jack D. Foner, *Blacks and the Military in American History: A New Perspective* (Praeger, 1974); Benjamin Quarles, *The Negro in the American Revolution* (University of North Carolina Press, 1961); Benjamin Quarles, *The Negro in the Civil War* (Little, Brown, 1953); and Dudley Taylor Cornish, *The Sable Arm: Negro Troops in the Union Army, 1861–1865* (Longmans, Green, 1956).
    2. Quarles, *The Negro in the American Revolution*, p. 14.
    3. Foner, *Blacks and the Military*, pp. 4–5.

A number of blacks (mostly in the South) joined the British forces against the American colonists, believing that a British victory would bring emancipation. In an effort both to prevent further defections by blacks and to deal with a critical shortage of colonial manpower, the Continental Congress allowed free black soldiers to reenlist. Without the formal approval of Congress, several states also recruited free blacks to fill their new draft quotas. Rhode Island, in desperate need of able-bodied fighters, even authorized the formation of an all-black battalion, the members of which were guaranteed freedom and equal pay and benefits.

Colonialists who were drafted to meet state levies could escape military service by supplying substitutes. The practice of using slaves as substitute draftees soon became popular in the North, as state restrictions on the use of black soldiers became even more relaxed. Only in the lower South did the continuing fear of slave revolts prevail over the exigencies of war and sustain provisions against the enlistment of blacks.

An estimated 5,000 blacks, including those with the Continental navy, the state navies, and privateers, fought with the colonial forces of the American Revolution.[4] Nevertheless, their services and achievements were quickly forgotten after the war. Political and social policy dictated that they be barred from the regular armed forces and militias of the new nation. Although blacks participated in a naval war with France and (against official policy) in the War of 1812, it was not until the Civil War that they were again formally allowed to bear arms.

The Civil War and Its Aftermath

During the early days of the Civil War, blacks were purposely excluded from service in the Union Army by the Lincoln administration in order to maintain the loyalty of border states and to focus the cause of the struggle on the preservation of the Union rather than the abolition of slavery. Some military leaders also feared that the presence of black soldiers in the army would create disharmony and drive away white volunteers. By the middle of 1862, however, white volunteers were becoming scarce and some black regiments were formed by Union generals without authorization.

The Emancipation Proclamation contained a provision for the enlistment of blacks, and soon after its issuance in 1863 active recruiting

---

4. Ibid., p. 15; Quarles, *The Negro in the American Revolution*, p. ix.

efforts were begun by the states. The first national draft law was passed two months later, and the states began to assemble volunteer black units whose enlistees could be counted in the states' draft quotas.

The nation's first draft law had another, more immediate effect on blacks. The Conscription Act of March 3, 1863, allowed a drafted man to hire a substitute or purchase his release from military service for a payment of $300. Many workingmen apparently believed that freed slaves would migrate north and usurp their jobs. Coupled with the popular view that the new draft law discriminated against the poor was an animosity toward blacks encouraged by certain antiwar factions: the draft was a result of the war; slavery was responsible for the war; and slaves were black.[5]

Thus soon after the names of the first Civil War draftees appeared in the New York City newspapers, riots broke out in several cities in the East and Midwest. The New York City riot was the most severe, causing an estimated 1,200 deaths and considerable property damage. Its tone was also that of an antiblack race riot, as several hundred blacks were killed and thousands fled the city. The streets were littered with the dead and dying, and the mutilated bodies of black victims hung from the trees and lampposts.[6]

Yet it was the participation of blacks in the Civil War, according to President Lincoln, that ensured a Northern victory and the preservation of the Union.[7] The Bureau of Colored Troops recruited and organized over 185,000 blacks into the U.S. Colored Troops.[8] Blacks accounted for about 9 to 10 percent of the Union Army and one-quarter of enlistments in the Navy (which officially authorized black enlistments in 1861).[9] When black volunteers in independent and state units are

---

5. Ulysses Lee, "The Draft and the Negro," *Current History,* vol. 55 (July 1968), p. 29.

6. See, among others, Harry A. Marmion, "Historical Background of Selective Service in the United States," in Roger W. Little, ed., *Selective Service and American Society* (Russell Sage Foundation, 1969), p. 37; William B. Hesseltine, *Lincoln and the War Governors* (Knopf, 1948), pp. 298–307; Quarles, *The Negro in the Civil War,* pp. 237–45; and Lee, "The Draft," p. 29.

7. Foner, *Blacks and the Military,* p. 48.

8. Lee, "The Draft," p. 30. Quarles *(The Negro in the Civil War,* p. 199) estimates that the number of black enlisted men in the Union Army was 178,975. By the end of the Civil War, there were 166 black regiments—145 infantry, 7 cavalry, 12 heavy artillery, 1 light artillery, and 1 engineering.

9. Quarles, *The Negro in the Civil War,* p. 230; and Foner, *Blacks and the Military,* p. 47.

included, it is estimated that close to 390,000 blacks served in the Civil War.[10]

More than 38,000 black soldiers lost their lives during that war—a mortality rate almost 40 percent higher than that of white troops. The largest number of deaths in any single outfit in the Union Army occurred in the Fifth United States Colored Heavy Artillery, where 829 soldiers died.[11]

After the Civil War, a congressional authorization created six black regiments in the regular Army (later reduced to two infantry and two cavalry regiments). These units, led by white officers, fought Indians and filled outposts in the West. "In a number of battles during the Indian Wars," according to one historian, "the four black regiments, especially the cavalry units, showed that in most situations they were the equal of white soldiers."[12] The Ninth and Tenth Cavalry also participated in the charge up San Juan Hill, and during the Mexican Punitive Expedition of 1916-17, "Gen. John Pershing designated the Tenth Cavalry to be the major part of his flying column in pursuit of Pancho Villa."[13] The existence of the regiments ensured that no blacks would serve in any other branch of the armed forces except in a national emergency. Civilian life held few opportunities for blacks, but in the post–Civil War Army they could find steady jobs and income, food and shelter, training in basic educational skills, and a degree of status. So these regiments seldom, if ever, had vacancies. During this period blacks made up about 10 percent of total Army strength.[14]

The status associated with service in the armed forces for blacks diminished greatly after the infamous "Brownsville Affray." In 1906 black soldiers stationed in Brownsville, Texas, allegedly rioted in protest

10. Lee, "The Draft," p. 30.
11. John Hope Franklin, *From Slavery to Freedom: A History of Negro Americans*, 5th ed. (Knopf, 1980), p. 224. Franklin states that the high mortality rate can be explained by several "unfavorable conditions," including "excessive fatigue details, poor equipment, bad medical care, the recklessness and haste with which Negroes were sent into battle, and the 'no quarter' policy with which Confederates fought them." Accounts vary on the number of blacks who died in the war. Cornish (*The Sable Arm*, p. 288), for example, observed that 68,178 black troops—more than a third of the total enrolled—lost their lives from all causes; of these, 2,751 were killed in action, and the rest died of wounds or disease or were reported missing.
12. Marvin Fletcher, *The Black Soldier and Officer in the United States Army, 1891–1917* (University of Missouri Press, 1974), p. 26.
13. Ibid., p. 154.
14. Foner, *Blacks and the Military*, pp. 52–55.

against their treatment by the townspeople. When the guilty men could not be found, President Theodore Roosevelt ordered the dishonorable discharge of three entire black companies without trial by court-martial. The 167 black soldiers—some with career-level service (up to twenty-seven years) and citations for bravery, and six with the Medal of Honor—were discharged without honor, back pay, allowances, benefits, pensions, or the chance to gain federal employment of any kind.[15]

Eleven years after the Brownsville episode, a more violent if less well-known incident involving the men of the all-black Twenty-fourth Infantry occurred in Houston, Texas. The white community of Houston, like the people of Brownsville, bitterly resented the presence of black troops (largely from northern states). Racial tension mounted, and there was an altercation between white policemen and black soldiers over the alleged abuse of a black woman. In retaliation, more than a hundred members of the Twenty-fourth Infantry mutinied against their officers, seized rifles and ammunition by force, and marched upon downtown Houston. Several policemen, citizens, and soldiers were killed, and many more were wounded.[16]

After the Houston race riot, the War Department indicted 118 soldiers and convicted all but 8 (who testified in return for immunity from prosecution) on charges of murder and mutiny. Thirteen men were "secretly" hanged—under the jurisdiction of the area commander and without the right of appeal to either the secretary of war or the president—in what was described by some as a "speedy execution" and a "military lynching" to placate the South. Another 6 soldiers were later hanged,

---

15. Fletcher, *The Black Soldier*, pp. 119–52; and Foner, *Blacks and the Military*, pp. 96–103. As Fletcher notes, 14 men were readmitted to military service after a year. In September 1972 the Department of the Army ordered the records of the other 153 black soldiers changed to honorable discharge, but with no award of back pay or allowances. In 1973 Congress granted the lone survivor of the Brownsville group a $25,000 pension along with medical benefits. Fletcher observes that "the Brownsville raid and the actions of the military before and after it point out quite clearly the negative aspects of the position of the black man in the service" (pp. 144,147,152).

16. Edgar A. Shuler, "The Houston Race Riot, 1917," in Allen D. Grimshaw, ed., *Racial Violence in the United States* (Aldine, 1969), pp. 73–87. In 1944 Gunnar Myrdal observed: "During his entire military history in the country, the Negro has experienced numerous humiliations of various kinds. He has been abused because of his race by many white officers, by white soldiers and by white civilians. There have been race riots in or around camps. The Negro soldier has usually been punished most severely when he was only one offender among many, and sometimes even when he was the victim." See *An American Dilemma: The Negro Problem and Modern Democracy* (Pantheon Books, 1972), vol. 1, p. 421.

and 63 were sentenced to life imprisonment. The rest were given dishonorable discharges and prison terms ranging from two to fifteen years.[17]

World War I

At the outbreak of World War I, blacks made up about 10.7 percent of the general population, and the Selective Service draft ensured that about that proportion served in the military. Ambrose notes that many blacks pinned their hopes for a better future on involvement in the war and many black leaders hoped to use the Army as a vehicle for social change. W. E. B. Dubois, for example, believed in 1917 that "if the black man could fight to defeat the Kaiser . . . he could later present a bill for payment due to a grateful white America."[18] But most black soldiers were draftees, since few were allowed to enlist; and most were assigned to traditional, menial occupations in peripheral units (supply, stevedore, engineer, and labor crews).

About 200,000 black soldiers served in France during the war; eight out of ten were assigned as laborers in the Service of Supplies. The relatively small number who served in combat units were subjected to unusually sharp criticism. In one of the most publicized incidents, the 368th Regiment of the 92nd "Buffalo" Division allegedly "became demoralized and fled to the rear during five days of the Meuse-Argonne offensive beginning September 26, 1918."[19] In contrast, the all-black 369th Infantry Regiment (*"Les Enfants Perdus"*), which served directly under the French, received high praise from its French commander:

17. Accounts differ about the number of people killed during the riot as well as the number of soldiers who were later court-martialed and sentenced to prison. This account is from Foner, *Blacks and the Military,* pp. 113–16. In 1918, Foner also notes, President Wilson reviewed the case and commuted the death sentences of ten (out of sixteen) soldiers. President Harding—after receiving a petition signed by 50,000 people—reduced the sentences of the men who were still in prison. In 1938 the last of the convicted Houston rioters was released from jail. As it turned out, this incident had an important impact on subsequent military jurisprudence. The manner in which it served as one of the catalysts for revision of the Articles of War and the Manual for Courts-Martial is outlined in Major Terry W. Brown, "The Crowder-Ansell Dispute: The Emergence of General Samuel T. Ansell," *Military Law Review,* vol. 35 (January 1967), pp. 1–45.

18. Stephen E. Ambrose, "Blacks in the Army in Two World Wars," in Stephen E. Ambrose and James A. Barber, Jr., eds., *The Military in American Society* (Free Press, 1972), pp. 178–79.

19. Lee Nichols, *Breakthrough on the Color Front* (Random House, 1954), p. 33. Many extenuating circumstances have been cited, such as inadequate training, inferior officers, and lack of firepower.

"They never lost a prisoner, a trench or a foot of ground during 191 days under fire, longer than any other American unit." And based on his first-hand observation, President Truman, who later acknowledged the problems of the 92nd Division, also assessed the performance of the all-black units that served under the French as "100 percent all right."[20]

The Navy allowed blacks to enlist in all service ratings at the start of the war. However, by 1918 blacks accounted for only about 1 percent of the naval forces—and most were messmen, stewards, or coal passers in the firerooms. In all, the Navy enlisted 10,000 blacks during World War I. The Marine Corps accepted no blacks.[21]

After the Armistice, the Navy began to recruit large numbers of Philippine nationals to fill messman vacancies, and virtually stopped enlisting black sailors. The few blacks accepted by the Navy (after 1932) were allowed to serve in the messmen's branch only. The Army remained segregated, and it adopted a policy of black quotas that would keep the number of blacks in the Army roughly proportionate to the number of blacks in the national population. The number in the Army never even approached the quota during this period—on the eve of World War II, the Army's own mobilization plan allowed for only about 6 percent of blacks in the total enlisted force.[22] No blacks were permitted to enlist in the Air Corps. And the Army's continuing commitment to a white officer corps was evidenced by the fact that there were only five black regular Army officers on duty, three of whom were chaplains.

World War II and Korea: The Fight for the Right to Fight

In the years just before World War II American blacks became increasingly concerned about the racial policies and conditions in the armed forces. Black leaders and organizations, leading black newspapers, concerned legislators, liberal coalitions, and others began to apply political pressure to break down the newly strengthened color barriers. The Selective Training and Service Act of 1940 reflected the efforts of

20. Ibid., p. 85. It has been reported, however, that a secret document was issued by General Pershing's headquarters urging French officers "not to treat Negroes with familiarity and indulgence, since this would affront Americans." Myrdal, *An American Dilemma,* vol. 1, p. 420.

21. Richard J. Stillman II, *Integration of the Negro in the U.S. Armed Forces* (Praeger, 1968), p. 16.

22. Richard M. Dalfiume, *Desegregation of the U.S. Armed Forces: Fighting on Two Fronts, 1939–1953* (University of Missouri Press, 1969), p. 23.

those who promoted "equality of service" by stipulating that the selection of volunteers and draftees for the armed forces should not discriminate against any person "on account of race or color."[23] However, since the act also stipulated that volunteers be "acceptable to the land or naval forces for such training and service," the separate military departments retained "unlimited discretion" to develop their own qualification standards for enlistment.[24]

World War II

Further pressure was brought to bear on President Franklin D. Roosevelt just before the election of 1940. The Roosevelt administration responded by enunciating the approved policy of the War Department, which included the following main points: (1) the proportion of blacks in the Army would be equivalent to the proportion of blacks in the general population; (2) black units would be established in each branch (combatant and noncombatant) of the Army; and (3) blacks would be allowed to attend officer candidate schools so they could serve as pilots in black aviation units. Yet the statement of policy also noted that, for the maintenance of troop morale and defense preparations, the War Department would continue "not to intermingle colored and white enlisted personnel in the same regimental organizations."[25] Furthermore, existing black units would receive no black reserve officers other than chaplains and medical officers.

The main focus of the "Negro problem" for the War Department between 1941 and 1943 was "ensuring that Negroes represented 10 percent of the Army, the same percentage that they composed of the general population"—a goal that was never reached.[26] Since segregation was a part of American life, the Army believed that it was a fixed part of the military establishment as well. The Army position was that the military should not be a laboratory for social experimentation; integration would hurt unit efficiency and create unnecessary racial friction. Black soldiers, because of the special treatment required, were thus viewed as

---

23. A complete documentation of racial policy during the World War II era can be found in Ulysses Lee, *The United States Army in World War II*, Special Studies: *The Employment of Negro Troops* (Office of the Chief of Military History, U.S. Army, 1966).

24. Ibid., pp. 71–74; Dalfiume, *Desegregation*, p. 31; Foner, *Blacks and the Military*, p. 137.

25. Dalfiume, *Desegregation*, p. 39.

26. Ibid., p. 44.

manpower problems rather than assets. A special study group, the Advisory Committee on Troop Policies, was eventually established in 1942 to investigate the extent of the race problem within the armed forces and to recommend appropriate action.[27]

The policy of racial segregation and quotas also created many unforeseen administrative difficulties during the early mobilization. All-black units, for example, had to be geographically placed so as to keep objections from local communities at a minimum. Also needed were special training staffs and separate facilities; separate assignment, classification, and replacement processes to segregate units and apportion blacks to different branches; and special procedures to identify men by race in order to limit the proportion of black draftees. The Selective Service System helped solve the problem of racial identification by creating different procedures and separate calls for the induction of blacks.

The Navy and the Marine Corps avoided the race issue entirely by accepting only white volunteers. This placed an added burden on the Army, which could not absorb more than its "fair share" of the nation's blacks. In 1942 the Navy relaxed its restrictions, and the Marine Corps enlisted blacks for the first time in its history. Eventually, the Navy and the Marine Corps were ordered to accept blacks through the draft, but difficulties in creating separate facilities and segregated units limited the number of blacks in these services.

In December 1944 shortages of infantry riflemen replacements in the European theater more or less compelled the Army to convert physically qualified men from the communication zone's all-black units into combat troops. A "call to arms" for black volunteers was made in the form of a circular letter, with instructions that it "be read confidentially to the troops . . . and made available in Orderly Rooms":

The Supreme Commander desires to destroy the enemy forces and end hostilities in this theater without delay. Every available weapon must be brought to bear upon the enemy. To this end the Commanding General, Com Z, is happy to offer to a limited number of colored troops who have had infantry training, the privilege of joining our veteran units at the front to deliver the knockout blow. . . .

The Commanding General makes a special appeal to you. It is planned to assign you without regard to color or race to the units where assistance is most needed, and give you the opportunity of fighting shoulder to shoulder to bring

---

27. Known as the McCloy Committee, the group was established as a "clearing house for staff ideas on the employment of Negro troops" and a liaison for the civilian community. Ibid., pp. 83, 86–88.

about victory. Your comrades at the front are anxious to share the glory of victory with you. Your relatives and friends everywhere have been urging that you be granted this privilege. The Supreme Commander, your Commanding General, and other veteran officers who have served with you are confident that many of you will take advantage of this opportunity and carry on in keeping with the glorious record of our colored troops in our former wars.[28]

Changes in the original plan directed that black volunteers be trained as members of separate infantry rifle platoons—available to Army commanders who would then provide leaders—so that they could be substituted for white units "in order that white units could be drawn out of line and rested." Yet the change in Army plans was never actually communicated to the black units during the period of volunteering. In two months 4,560 black soldiers had volunteered, some taking reductions in rank for the privilege of signing on. Eventually the flow of volunteers was halted, as 2,800 black troops were hurriedly retrained as infantrymen and sent into action.[29]

The Army sent the newly trained black volunteers (approximately fifty platoons) to fight alongside white troops in France, Belgium, and Germany. As the Battle of the Bulge intensified, all-black platoons were combined with white platoons and put into action as elements of eleven divisions of the U.S. First and Seventh armies. Division commanders reportedly were "delighted" with the performance of the Negro platoons; reflecting the spirit of the venture, the black platoons assigned to the famed "Timberwolf" Division called themselves the "Black Timberwolves," contending that they were the "fiercest of all Timberwolves."[30]

Disagreement about the fighting abilities of blacks arose again during World War II. In February 1945 a task force of the 92nd "Buffalo" Division that had been criticized in World War I was pulled out of action against the German Gothic Line in Italy after three days of excessive "straggling" and "disorganization."[31] Responding to accounts that the

---

28. The original letter was later modified because it was felt that the plan to use black troops constituted an unnecessarily radical break with traditional Army policy and existing regulations; further, it might prove embarrassing to the War Department. A revised letter was therefore prepared, changing all but the first two sentences of the original and no longer promising "the opportunity of fighting shoulder to shoulder." A cover memorandum also ordered the return and destruction of all copies of the original version—but by the time the revised letter and new orders were released, the first letter had already been distributed to most of the units. See Lee, *Employment of Negro Troops*, pp. 689–91.
29. Ibid., pp. 691–95.
30. Nichols, *Breakthrough on the Color Front*, p. 69.
31. Ibid., p. 16.

92nd was "melting away" in the heat of battle, Truman Gibson, a civilian aide to the secretary of war, visited the division in Europe and reported:

It is a fact that there have been many withdrawals by panic-stricken infantrymen. However, it is equally evident that the underlying reasons are quite generally unknown in the division. The blanket generalizations expressed by many, based on inherent racial difficulties, are contradicted by many acts of individual and group bravery.[32]

In covering the story, the white press emphasized Gibson's acknowledgment that the 92nd had been subject to panicky retreats and his reports of low literacy among black troops.[33] The black press was generally harsh in its criticism of Gibson, who was black, and some newspapers called for his resignation.[34] Despite the adverse publicity at home, the 92nd was awarded the Cross for Merit of War by the Italian government in 1945.[35]

A board of officers (the Special Board on Negro Manpower headed by Lieutenant General Alvan Gillem), established to evaluate the performance of black soldiers, concluded that "all-Negro divisions gave the poorest performance of Negro troops," but spoke favorably of "the performance of Negro Infantry platoons fighting in white companies." The poor showing of all-black units, according to the Gillem report, was in part the result of the Army's poor preparation and planning.[36] Most assessments of the performance of black troops skirted the question of the quality of their leaders, most of whom were white. But there is ample evidence to suggest that at least some of the blame for the poor record of black units should rest on inferior leadership. Black units "often became, as they had in earlier wars, dumping grounds for officers unwanted in white units."[37] Moreover, many white officers resented

---

32. Lee, *Employment of Negro Troops*, p. 576.

33. Ibid., p. 577. As of March 1945, 75 percent of the men in the division had scored below the thirtieth percentile on the standardized aptitude test.

34. Richard O. Hope, *Racial Strife in the U.S. Military: Toward the Elimination of Discrimination* (Praeger, 1979), p. 29.

35. Department of the Army, General Order 43, December 19, 1950. Since the order did not include the usual citation, the circumstances surrounding the award are unclear. It is known that, following the aborted February thrust, two black regiments in the division were consolidated into the third and were replaced by two others: the 473rd Infantry, composed of battle-hardened white veterans of the African campaign, and the 442nd Infantry, the Japanese-American fighting team that had already made a name for itself in southern France. In April 1945 the reconstituted division mounted offensive operations along the Ligurian coast. Nichols, *Breakthrough on the Color Front*, p. 17.

36. Samuel A. Stouffer and others, *The American Soldier: Adjustment During Army Life*, vol. 1 (Princeton University Press, 1949), p. 586.

37. Morris J. MacGregor, Jr., *Integration of the Armed Forces, 1940–1965* (U.S.

being assigned to a black unit, viewing it as a stigma and a road to nowhere. The Army aggravated the situation "by showing a preference for officers of southern birth and training," who were particularly resented by the black troops.[38] A 1942 survey of newly commissioned officers serving in black units revealed that "each of them had been disappointed on learning of his assignment . . . that they had failed to measure up and thought that they were assigned to inferior service."[39]

In his assessment of the performance of black soldiers in World War II, Eli Ginzberg wrote:

While many Negroes saw military service as an opportunity to prove their individual worth and to help raise the prestige of their group, thereby striking a blow against segregation, many others failed to do their best. They could not free themselves from the crippling experiences which had been theirs from earliest childhood—"as a man is treated, so he is likely to respond." The Army in turn was greatly handicapped in making effective use of Negro manpower. Segregation interfered with the optimal training and assignment of Negroes with high potential; it led to a serious imbalance of skills and aptitudes in Negro divisions; and it was reflected in serious weaknesses in the leadership of Negro units.

In the face of the handicaps which they brought with them into the Army and the barriers which they encountered once they were in uniform, the remarkable finding is that the vast majority of Negroes performed satisfactorily, not that they accounted for a disproportionate number of ineffectives.[40]

The chief historian of the Army, Walter L. Wright, Jr., made a somewhat similar observation just before the close of the war in 1945:

American Negro troops are, as you know, ill-educated on the average and often illiterate; they lack self-respect, self-confidence, and initiative; they tend to be very conscious of their low standing in the eyes of the white population and consequently feel very little motive for aggressive fighting. In fact, their survival as individuals and as a people has often depended on their ability to subdue completely even the appearance of aggressiveness. After all, when a man knows that the color of his skin will automatically disqualify him for reaping the fruits of attainment it is no wonder that he sees little point in trying very hard to excel

---

Army, Center of Military History, 1981), p. 37. George Custer, it is said, turned down the colonelcy in a black regiment for a lower-ranking position in a white regiment. Fletcher, *The Black Soldier,* p. 21. There were some clear exceptions. For example, the task force of the 92nd that was pulled out of action in 1945 was commanded by Lieutenant Colonel Edward L. Rowny, a West Point graduate, who became a general officer and, after his retirement, an ambassador.

38. MacGregor, *Integration of the Armed Forces,* p. 37.
39. Lee, *Employment of Negro Troops,* p. 185.
40. Eli Ginzberg and others, *The Ineffective Soldier: Lessons for Management and the Nation,* vol. 1: *The Lost Divisions* (Columbia University Press, 1959), pp. 124–25.

anybody else. To me, the most extraordinary thing is that such people continue trying at all.

"The conclusion which I reach is obvious," Wright continued. "We cannot expect to make first-class soldiers out of second or third or fourth class citizens. The man who is lowest down in civilian life is practically certain to be lowest down as a soldier. Accordingly, we must expect depressed minorities to perform much less effectively than the average of other groups in the population."[41]

Of the more than 2.5 million blacks who registered for the draft in World War II, about 909,000 served in the Army. In 1944 there were over 700,000 blacks in the Army; this represented the greatest proportion of blacks to total Army strength in World War II. So at its peak, only 8.7 percent of the Army—instead of the planned 10 percent—was black. In June 1945 blacks accounted for less than 3 percent of all men assigned to combat duty in the Army. About 78 percent of all black males—and only 40 percent of all white males—in the Army were placed in the service branches (including quartermaster, engineer, and transportation corps).[42]

Approximately 167,000 blacks served in the Navy during the war, about 4 percent of total Navy strength; and over 17,000 blacks enlisted in the Marine Corps, 2.5 percent of all marines.[43]

"Despite the multitude of problems with which the Army was faced in the use of Negro troops in World War II," historian Ulysses Lee would later write in the Army's official account of the war, "at the war's end a greater variety of experience existed than had ever before been available within the American Military Establishment":

> They had been used by more branches and in a greater variety of units, ranging from divisions to platoons in size and from fighter units to quartermaster service companies in the complexity of duties. They had been used in a wider range of geographical, cultural, and climatic conditions than was believed possible in 1942. All of this was true of white troops as well, but in its manpower deliberations and in its attempts to wrest maximum efficiency and production from the manpower allotted to it, the Army found that it was the 10 percent of

---

41. Lee, *Employment of Negro Troops*, pp. 704–05.
42. H. S. Milton, ed., *The Utilization of Negro Manpower in the Army*, Report ORO-R-11 (Chevy Chase, Md.: Operations Research Office, Johns Hopkins University, 1955), p. 562; and Ambrose, "Blacks in the Army," p. 186.
43. Data on the number of blacks serving in the Navy and Marine Corps are from Foner, *Blacks and the Military*, pp. 172–73. Proportions were computed on the basis of historical statistics found in Department of Defense, *Selected Manpower Statistics* (Directorate for Information, Operations, and Reports, 1972), pp. 83–85.

American manpower which was Negro that spelled a large part of the difference between the full and wasteful employment of available American manpower of military age.[44]

The Postwar Period

After the war the Army was faced with the prospect of a greatly increased proportion of blacks. Many black soldiers wanted to remain in the Army, in spite of any unfair treatment, since the military offered much better opportunities than could be found in civilian life. And many white soldiers wanted to be discharged as quickly as possible.[45]

Increasing pressure and protest were focused by the black community on the Truman administration and the armed forces for changes in policy toward black service members. The United States, after all, had maintained a segregated military throughout its struggle against a nation that preached a master race ideology. At a time when the nation stood united against fascism and claimed to be the last great fortress of democracy, it still preserved the racist policies and practices of its military institutions.[46] American blacks, it is said, took advantage of the war to point out the differences between the American creed, for which the war was being fought, and actual practice. Many blacks, stimulated by the rhetoric and ideological aims of the war, found cause to reexamine their "place" in American society.[47]

It has been observed that the right to bear arms is an integral part of the normative definition of citizenship.[48] Political rights are to be achieved by participation in the military and by proof of loyalty through defense of the state. Indeed, the absorption of immigrants into the American melting pot has been achieved historically through the "blood test"— you proved you loved America through allegiance and sacrifice and dying for the country in its wars (that is, paying the "price in blood").[49]

---

44. Lee, *Employment of Negro Troops*, pp. 703–04.
45. Milton, *Utilization of Negro Manpower*, p. 562.
46. See Dalfiume, *Desegregation*, p. 107; and Foner, *Blacks and the Military*, p. 134.
47. See "The Negro's Mind and Morale" (chapter 6) in Dalfiume, *Desegregation*, pp. 105–31.
48. Mark Jan Eitelberg, "Military Representation: The Theoretical and Practical Implications of Population Representation in the American Armed Forces" (Ph.D. dissertation, New York University, 1979), pp. 71–88, 255–57.
49. As Michael Novak observes in *The Rise of the Unmeltable Ethnics: Politics and Culture in the Seventies* (Macmillan, 1972), pp. xxi–xxii, when the Poles were only about 4 percent of the U.S. population in 1917–19, they accounted for over 12 percent of the nation's casualties in World War I. The "fighting Irish" did not win that epithet on the

It is also assumed that it is the obligation of citizens to serve their nation and to participate in the armed forces whenever necessary. But members of social categories have been denied civil rights in the past on the grounds that their capacity to fulfill the military obligation was restricted.[50] Thus the military establishment that excludes special groups from equal service imposes on them the overt stigma of civic inferiority. "Equality of service" in the American military was, for many blacks, the chance to shed existing stereotypes of racial inferiority and second-class citizenship. Throughout the war and its aftermath, the fight for blacks was for the "right to fight."

On July 26, 1948, just three months before the presidential election, President Harry S. Truman issued an executive order, which "declared to be the policy of the President that there shall be equality of treatment and opportunity for all persons in the armed services without regard to race, color, religion, or national origin," and that promotions were to be based "solely on merit and fitness." The order also established the President's Committee on Equality of Treatment and Opportunity (whose chairman was Charles H. Fahy) to work with the secretary of defense and the service secretaries in implementing the new policy.[51] Executive Order 9981 "shook the Defense Department to its foundations" and raised the ire of segregationists in Congress.[52]

---

playing fields of Notre Dame but by dying in droves during the American Civil War. Victor Hicken points out in *The American Fighting Man* (Macmillan, 1969), p. 365, that the sansei, because of Pearl Harbor and subsequent discrimination, felt compelled to prove their loyalty to America on the battlefield "with a vengeance." Thus, according to Morris Janowitz, "from World War I onward, citizen military service has been seen as a device by which excluded segments of society could achieve political legitimacy and rights"; "Military Institutions and Citizenship in Western Societies," *Armed Forces and Society*, vol. 2 (Winter 1976), p. 192.

50. This was the case, for example, in the Dred Scott majority opinion of the Supreme Court in 1857. The Supreme Court ruled that the slave Dred Scott was not a citizen of the United States because (among other reasons) he was not a legitimate or equally obligated participant in the American armed forces.

51. Executive Order 9981, *Federal Register*, vol. 13 (July 28, 1948), p. 4313. See also *Freedom to Serve: Equality of Treatment and Opportunity in the Armed Services*, Report by the President's Committee (Government Printing Office, 1950).

52. *The Power of the Pentagon* (Washington, D.C.: Congressional Quarterly, 1972), p. 34. Even before the president delivered the executive order, there was strong opposition. Reflecting the typical views of segregationists, Senator Richard B. Russell of Georgia (later chairman of the Senate Committee on Armed Services), for instance, spoke on the Senate floor of how "the mandatory intermingling of the races throughout the services will be a terrific blow to the efficiency and fighting power of the armed services. . . . It is sure to increase the numbers of men who will be disabled through communicable diseases. It will increase the rate of crime committed by servicemen"; ibid., pp. 34–35. Note too

When the Fahy Committee met for the first time in early 1949, it discovered that the Navy had already claimed some progress toward racial equality with the establishment of its own integration and nondiscrimination policy in 1946. The Marine Corps had abolished segregation in basic training but still maintained all-black units. The Air Force favored integration and was prepared to promote a new policy that would end racial quotas, open all occupational specialties, and base promotion on personal merit and abilities alone. The Army, with 40 percent of its occupational specialties and 80 percent of its training schools without black soldiers, was firmly opposed to any changes in its traditional racial policy.[53] The resistance of the Army leadership gained some strength from the conclusions of several studies on black manpower. The Gillem Board, for example, recommended that no fundamental changes be made in the Army's policy of segregation and its use of racial quotas.[54] In 1950 an Army board similarly found that widespread integration (however desirable as a social measure) and abolition of the 10 percent ceiling on black manpower would markedly reduce combat efficiency and unit morale.[55] The Army further contended that its racial policies were not

---

that Executive Order 9981 did not specifically promise integration; it promised "equality of treatment and opportunity." Since the policy of the armed forces was "separate but equal" treatment of the races, the order was the subject of some controversy and confusion and not a little evasion when first issued.

53. "Equality in the Military: 25-Year Progress Report," *New York Times*, May 30, 1975.

54. The report of the Special Board on Negro Manpower (*Policy for Utilization of Negro Manpower in the Post-War Army: With Recommendations for Development of Means Required and a Plan for Implementation of the Same,* November 1945, and the supplemental report of January 1946) is summarized and evaluated in Milton, *Utilization of Negro Manpower,* pp. 574–79. Even though the Gillem Board did not question the Army's traditional policy of segregation, its recommendations were in contrast to the findings of earlier study groups and much more favorable to the development of improved racial policies. For example, the recommendation for enlistment of blacks in proportion to their representation in the general population actually encouraged a greater peacetime utilization of black manpower than the Army had previously envisioned. In fact, the Gillem Board found that during World War II "adequate plans were not prepared for the ultimate utilization of this [black civilians available for military service] manpower." Furthermore, the board recognized the rights and responsibilities of blacks as citizens to participate in the military and recommended the adoption of "a progressive policy for greater utilization of the Negro manpower." The armed forces "must eliminate, at the earliest practicable moment, any special consideration based on race." See Alan L. Gropman, *The Air Force Integrates, 1945–1964* (Office of Air Force History, 1978), pp. 46–56.

55. The report of the Chamberlin Board (Board to Study the Utilization of Negro Manpower) is examined in Milton, *Utilization of Negro Manpower,* pp. 579–81. Also reviewed are other supporting studies by the Army in the early 1950s on the limited use of black troops—including officer surveys and reports, Army student reports, Army com-

dictated by racial prejudice, but by two conditions: most whites do not associate with blacks, and blacks, through no fault of their own, do not have the skills or education required for many of the Army's occupational specialties.

The Fahy Committee, noting the difference between black and white soldiers in education and mental achievement, urged the Army to substitute an achievement quota for its racial quota. The Army, it was pointed out, could adjust its General Classification Test minimum qualification scores up or down and use its physical, psychiatric, and moral standards to effectively regulate the number of black enlistments. The Army could also make it difficult for soldiers to reenlist if they were "perennial low-score men or otherwise inapt."[56]

Without the committee's knowledge, President Truman made an informal agreement with the Army. "If, as a result of a fair trial of this new system, there ensues a disproportionate balance of racial strengths in the Army," Secretary of the Army Gordon Gray reaffirmed in a letter to President Truman, "it is my understanding that I have your authority to return to a system which will, in effect, control enlistments by race."[57] With the promise that a racial quota could be reinstated if necessary, the Army in 1950 became the last service to officially set forth a plan for unrestricted "equality of treatment and opportunity."

The Korean War

Despite the desegregation order, the outbreak of the Korean War found a still-segregated Army, with the all-black 24th Infantry Regiment committed to combat duty. Reports soon circulated about the unreliability of black soldiers, who allegedly would "melt into the night" only

---

mittee studies, and attitude surveys of black and white soldiers. After the Fahy Committee issued its report, the Chamberlin Board was asked to "reevaluate" its conclusions based on the Korean War experience. The board concluded that integrated combat units performed better than segregated units, but that it was necessary to reimpose the quota and to retain some separate black units. The second Chamberlin report was not approved.

56. Memorandum to the president from David K. Niles, February 7, 1950, and supporting documents, in Morris J. MacGregor and Bernard C. Nalty, *Blacks in the United States Armed Forces: Basic Documents, Volume XI: Fahy Committee* (Wilmington, Del.: Scholarly Resources, Inc., 1977), pp. 1343–45.

57. Letter to the president from Secretary of the Army Gray, March 1, 1950, in ibid., p. 1350. See also "Equality in the Military: 25-Year Progress Report"; and Dalfiume, *Desegregation*, pp. 197–98.

to turn up the next day insisting they had been lost.[58] But the charges subsided as the growing need for combat troops led to ad hoc integration, which by the war's end had become standard practice, largely as a result of efforts by General Matthew Ridgway.[59]

Blacks enlisted in large numbers; by the middle of 1951 one out of every four new recruits in the Army was black.[60] Black training units in the United States and service units in Korea could no longer absorb the rapidly increasing number of black enlistees. At the same time, the Army was faced with a shortage of men in white units, especially those on the front lines in Korea. Military necessity therefore more or less forced integration of both training units and combat units. Soon all Army basic training centers were integrated and blacks were being assigned to white combat units.

The Korean War, because of its coincidence with the new racial policy of the armed forces, also afforded an unusual opportunity to test the effectiveness of integration on a large scale. The most notable research undertaking of the period, "Project Clear," concluded in 1951 that "racial segregation limits the effectiveness of the Army" and, conversely, that "integration enhances the effectiveness of the Army." Black soldiers integrated into previously all-white combat units received high marks, particularly when contrasted with their counterparts in all-black units. In fact, social scientists found little difference in the fighting abilities of blacks and whites.[61] As matters turned out, "the Korean

58. Nichols, *Breakthrough on the Color Front*, p. 20. By one account, a song, called "The Bug-Out Boogie," partly self-deprecatory but also derogatory of white officers, was attributed to the 24th Infantry Regiment.

> When them Chinese mortars begin to thud
> The old Deuce-Four begin to bug.
> When they started falling 'round the CP tent
> Everybody wonder where the high brass went.
> They were buggin'out
> Just movin' on.

See Charles C. Moskos, Jr., *The American Enlisted Man: The Rank and File in Today's Military* (Russell Sage Foundation, 1970), pp. 239–40, note 4. "The incident of the song," according to one observer, "points up one of the extremely thorny problems that has plagued the Army in this war—the problem of how to get the most efficient combat use out of the nation's vast resources of Negro manpower; how best to employ the Negro's capacity to endure hardship, his high good humor that makes for high morale." Harold H. Martin, "How Do Our Negro Troops Measure Up?" *Saturday Evening Post*, June 16, 1951, p. 31.

59. Nichols, *Breakthrough on the Color Front*, p. 116.

60. Milton, *Utilization of Negro Manpower*, p. 569.

61. Ibid.; and Leo Bogart, ed., *Social Research and the Desegregation of the U.S. Army: Two Original 1951 Field Reports* (Markham, 1969).

conflict was the *coup de grâce* for segregation in the Army,"[62] and put to rest, at least on the surface, not only doubts about the individual effectiveness of black soldiers, but also fears that integration would have adverse consequences for group solidarity and hence unit performance. The Korean experience consistently supported the racial contact hypothesis developed during World War II: the more contact white soldiers had with black troops, the more favorable their attitude toward racial integration was.[63]

Thus Project Clear scientists recommended that "the Army should commit itself to a policy of integration to be carried out as rapidly as operational efficiency permits."[64] These findings, especially, bolstered proponents of "equality of service" and helped speed up changes in racial policy. By the end of the Korean War, more than 90 percent of all blacks in the Army were assigned to integrated units. The Air Force and the Marine Corps had eliminated their racially segregated units. And "young Negro recruits serving in Korea found it hard to believe that an all-Negro infantry regiment had ever existed."[65]

After Korea: The Halcyon Years

After the armistice, it did not take long for the military to complete its desegregation program. On October 30, 1954, the Pentagon announced that all-Negro units had been abolished,[66] bringing to an end "a quiet racial revolution . . . with practically no violence, bloodshed, or conflict."[67] But since the military had moved so far ahead of society, Army

---

62. Moskos, *The American Enlisted Man*, p. 111.
63. See Stouffer and others, *The American Soldier*, p. 594.
64. Milton, *Utilization of Negro Manpower*, pp. 5–6. The Project Clear report and supporting material, based on Korean and U.S. research studies in 1951, were originally classified "Confidential" and downgraded to "For Official Use Only" in 1963.
65. Lee, "The Draft," p. 33. However, integration and equality of service in the Navy at this point still moved "under easy sail." Until 1944, Stillman notes in *Integration of the Negro* (p. 5), all of the Navy's 165,000 blacks were assigned to the stewards' branch (where blacks were nicknamed "cooks and bellhops at sea"). Even though the Navy ended World War II with "the most progressive Negro policy of all the armed services," a great difference between policy and practice existed during the postwar years; see Dalfiume, *Desegregation*, p. 103. In March 1953 there were approximately 23,000 blacks in the Navy; almost half were in the stewards' branch. Renewed political pressure encouraged the Navy to open its occupational specialty groups to all seamen and to discontinue its separate recruitment of stewards one year later; see Foner, *Blacks and the Military*, p. 193.
66. "Services Abolish All-Negro Units," *New York Times*, October 31, 1954.
67. Dalfiume, *Desegregation*, p. 219.

posts became "islands of integration in a sea of Jim Crow."[68] The Pentagon, in fact, acknowledged that much remained to be done in military-community relations: "It is paradoxical that the Negro citizen in uniform has frequently been made to feel more at home overseas than in his home town."[69]

The major problems that blacks in the armed services confronted during the late 1950s stemmed from the racism that prevailed in many communities surrounding military installations.[70] Black service members not only faced the hostility of many civilians, but had difficulty finding decent living accommodations, restaurants, and schools. While the Pentagon was acutely aware of off-base discrimination, there was "no evidence that the Department of Defense ever worked for blacks off the post before the 1960s."[71]

Meanwhile, race relations within the military appeared to be relatively tranquil, particularly in contrast to the rest of society.

Overt expressions of prejudice in public are a rarity. Facilities are shared with no color line. For most white personnel the Army is their first experience of close contact for a prolonged time with a large group of Negroes. Negro-white friendships based on equality are formed. A degree of resentment is held by some individuals but in no sense is the Army sitting on top of a racial volcano. . . . Opinions concerning individual Negroes are closely correlated with that Negro's work habits, personality and intelligence rather than his race per se.[72]

It was not until the Kennedy era that the Pentagon took a more active role in dealing with off-base discrimination. One of the first steps was the reactivation in 1962 of the President's Committee on Equal Opportunity in the Armed Forces, called the Gesell Committee after its chairman, attorney Gerhard A. Gesell. The committee examined the "special efforts" and methods "to increase the presently insufficient flow of qualified Negroes into the Armed Forces" and the various factors that may have accounted for the fact that "participation of the Negro in the Armed Forces is less than the percentage of Negroes in our total population."[73] The committee found an unbalanced grade distribution

68. Pfc. Charles C. Moskos, Jr., "Has the Army Killed Jim Crow?" *Negro History Bulletin*, vol. 21 (November 1957), p. 29.

69. *New York Times*, October 31, 1954.

70. For an account of the Air Force's problems in integrating its units at several bases, see Gropman, *The Air Force Integrates*, especially chap. 4.

71. Ibid., p. 155.

72. Moskos, "Has the Army Killed Jim Crow?" p. 29.

73. President's Committee on Equal Opportunity in the Armed Forces, *Equality of Treatment and Opportunity for Negro Military Personnel Stationed Within the United States: Initial Report* (GPO, 1963), pp. 5, 12, 14.

of blacks in the services, segregation (or only token integration) and exclusionary practices in the National Guard and the reserves, and racial discrimination on military installations and in surrounding communities. The Gesell Committee report was invaluable because it presented for the first time since the Truman administration a detailed, quantitative picture of the relationship between blacks and the military.[74]

Vietnam and the New Era of Racial Representation

When the Gesell Committee submitted its final report to the president in 1964, there were about 860,000 enlisted personnel in the Army. In just one year the nation's military was committed to another war; and by 1968 the Army's enlisted force had expanded to include 1.4 million men and women. Institutionalized racial discrimination once more became a subject of both interest and controversy. However, in contrast to the two World Wars and the early days of Korea when blacks had to "fight for the right to fight," the advent of the Vietnam War brought charges that blacks were doing more than their fair share of the fighting. And unlike the Gesell Committee, many black leaders and others now questioned the "special efforts" and methods that *favored* the recruitment of blacks over whites.

Meanwhile, popular magazines reported official Department of Defense statistics showing that blacks were more likely to (1) be drafted, (2) be sent to Vietnam, (3) serve in high-risk combat units, and consequently (4) be killed or wounded in battle.[75] Between 1961 and 1966, when blacks composed approximately 11 percent of the general population aged nineteen to twenty-one, black casualties amounted to almost one-fourth of all losses of Army enlisted personnel in Vietnam.[76]

Protest movements soon converged on the machinery of the draft. Before Vietnam, the Selective Service System had operated in an

---

74. Stillman, *Integration of the Negro*, p. 110. The Gesell Committee's findings were also instrumental in the creation of the Office of the Deputy Assistant Secretary of Defense for Civil Rights and Industrial Relations.

75. See, for example, "The Draft: The Unjust vs. the Unwilling," *Newsweek*, April 11, 1966, pp. 30–32, 34; "How Negro Americans Perform in Vietnam," *U.S. News & World Report*, August 15, 1966, pp. 60–63; "Democracy in a Foxhole," *Time*, May 26, 1967, pp. 15–19; "King Talk," *National Review*, April 18, 1967, pp. 395–96; "Negroes Go To War," *The Economist*, April 15, 1967, p. 255; Karl H. Purnell, "The Negro in Vietnam," *The Nation*, July 3, 1967, pp. 8–10; "The Negro and Vietnam," *The Nation*, July 17, 1967, pp. 37–38; and "Negroes in the Vietnam War," *America*, June 10, 1967, pp. 827–28.

76. "How Negro Americans Perform in Vietnam," pp. 60–64.

environment of public and congressional approval; Selective Service could point to the general lack of public protest as proof that inequities, though they might exist, were not strongly felt.[77] For years, Selective Service had spent more time designing deferments than obtaining inductions. Now, "channeling" and the full array of deferments and disqualifications offered numerous ways for young men to avoid the draft.

The final report of the National Advisory Commission on Selective Service (the Marshall Commission) buttressed the charges of institutional racism made by black leaders. The commission gave "careful study to the effect of the draft on and its fairness to the Negro." It found that in October 1966 only 1.3 percent of all local draft board members were black, and in seven states local draft boards had no black members at all. The commission concluded that "social and economic injustices in the society itself are at the root of inequities which exist," and recommended that local draft boards "should represent all elements of the public they serve."[78]

Oddly enough, at about the time the Selective Service System was being denounced for favoring the induction of blacks and the poor, a liberal faction was criticizing the armed forces for systematically excluding blacks and the least-educated and least-mobile young men. Daniel P. Moynihan wrote:

History may record that the single most important psychological event in race relations in the 1960's was the appearance of Negro fighting men on the TV screens of America. Acquiring a reputation for military valor is one of the oldest known routes to social equality. . . . Moreover, as employment pure and simple, the armed forces have much to offer men with the limited current options of,

---

77. This observation was made by James M. Gerhardt, *The Draft and Public Policy: Issues in Military Manpower Procurement, 1945–1970* (Ohio State University Press, 1971), p. 361.

78. *In Pursuit of Equity: Who Serves When Not All Serve?* Report of the National Advisory Commission on Selective Service (GPO, 1967), pp. 9–10. There were also charges by civil rights leaders and others of blatantly "abusive discrimination against black registrants" by local draft boards. "White draft officials," some contended, "are using the power of the draft to punish Negroes." The draft and the Selective Service System were thus viewed as an "instrument of American racism." See Jean Carper, *Bitter Greetings: The Scandal of the Military Draft* (Grossman, 1967), pp. 144–45. The report of the Marshall Commission led to a second study (by the Task Force on the Structure of the Selective Service System), which disagreed emphatically with the Marshall Commission study and concluded that the draft system was overall highly satisfactory. A third study (by the Civilian Advisory Panel on Military Manpower Procurement, or the Clark Panel) also supported the basic organizational philosophy of the Selective Service System and rejected the conclusion that student and occupational deferments were inequitable.

say, Southern Negroes. By rights, Negroes are entitled to a *larger share* of employment in the armed forces and might well be demanding one.[79]

Moynihan's basic contention in 1966 was that the American armed forces had become "an immensely potent instrument for education and occupational mobility," but because of certain mental and physical requirements (perhaps overstated acceptance standards), "a whole generation of poor Negroes and whites are missing their chance to get in touch with American society."[80] Moynihan used as evidence the fact that blacks, high school dropouts, the unskilled, and the poor—a profile of poverty in the 1960s—were those most likely to be rejected by the services.

At the time, the Selective Service System was far from equipped to run employment or rehabilitation programs for the disadvantaged, and the Defense Department was reluctant to enter the "social welfare business." But the Great Society drafted the armed forces to help fight the War on Poverty, and "Project One Hundred Thousand" was launched. Intended to rehabilitate the nation's "subterranean poor," Project One Hundred Thousand was an experimental program for the annual induction of 100,000 men who would ordinarily be screened out primarily because of limited educational background or low educational attainment.

Between October 1966 and June 1969, approximately 246,000 recruits entered the service under the program. Forty percent of these men were black (the control group—other male enlistees during the same period—was about 9 percent black); almost 50 percent were from the South (almost 28 percent in the control group); and 47 percent were draftees. Unfortunately, not many of the recruits could qualify for the "hard skill" occupations that would help them in civilian life. More incoming personnel were being processed by the armed forces than in the past fifteen years, and automated processing methods were used to assign new recruits. There were no special placement programs yet, so most of the "new standards" men found themselves, as a matter of ordinary procedure, in the "soft skills" or "simpler jobs." Consequently, about 37 percent of Project One Hundred Thousand recruits were assigned to combat-type skills, and over half who entered the Army and the Marine Corps were sent to Vietnam.[81]

79. Daniel P. Moynihan, "Who Gets in the Army?" *New Republic,* November 5, 1966, p. 22; emphasis added.
80. Ibid., pp. 20, 22.
81. *Project One Hundred Thousand: Characteristics and Performance of "New*

To the black community the most distressing aspect of Selective Service inequities and overstated entry standards was that the armed forces were apparently sending its "best" young men—those who were educated and healthy but not deferred—to fight in Vietnam. The majority of blacks who applied to the military (conscripts or volunteers) were rejected because of inadequate education or poor health. Those who were being accepted, according to Whitney Young, were the "cream of the crop" from the black community—the "potential forces of leadership . . . in the battle cry for freedom at home."[82] They were the young, income- and family-producing males,[83] the potential black leaders who were leaving a leadership vacuum in the black community that could not be filled by appropriate substitutes.[84] And worst of all, as Moskos points out, the economic and educational disadvantages that made blacks *available* for military service made them at the same time *unavailable* for many technical job opportunities in expanding skill areas within the armed forces.[85]

Studies of the Vietnam-era draft and casualty data have generally disputed the existence of widespread institutional racism. More precisely, evidence points to an institutionalized bias based on social or economic class. In 1968 Davis and Dolbeare examined the draft mechanism and concluded that "there is little evidential basis for doubting the existence of economic discrimination in deferment/induction policies."[86] They found a definite income-based pattern of military service, with the incidence of military service occurring most often in the lower-middle socioeconomic bracket. They also found that blacks were overrepresented among draftees, not as a function of their race, but as a function of their economic status (the disproportionate numbers of them in the low-income strata); and although there was a high incidence of rejections

---

Standards" Men (Office of the Assistant Secretary of Defense for Manpower and Reserve Affairs, 1969). See also Lawrence M. Baskir and William A. Strauss, *Chance and Circumstance: The Draft, the War and the Vietnam Generation* (Knopf, 1978), pp. 126–31.

82. Whitney M. Young, Jr., "When the Negroes in Vietnam Come Home," *Harper's,* June 1967, p. 66.

83. See Robert D. Tollison, "Racial Balance and the Volunteer Army," in Miller, *Why the Draft?* p. 149.

84. Harry A. Marmion, *The Case Against a Volunteer Army* (Quadrangle Books, 1971), p. 34.

85. Charles C. Moskos, Jr., "Minority Groups and Military Organization," in Ambrose and Barber, *The Military in American Society,* p. 195.

86. James W. Davis, Jr., and Kenneth M. Dolbeare, *Little Groups of Neighbors: The Selective Service System* (Markham, 1968), p. 129.

of the poor (particularly the black poor), those who passed their physical and mental examinations were more likely to be drafted than men with similar qualifications but higher incomes.[87]

Incidents with racial overtones plagued the Vietnam period. Among the most widely publicized were a race riot among prisoners in a stockade in Vietnam in 1968[88] and several incidents aboard naval vessels in the early 1970s.[89] Even the Air Force, which had been virtually free of racial problems, did not escape. Four days of rioting in May 1971 at Travis Air Force Base, California, were ignited by racial incidents on the base; 110 blacks and 25 whites were arrested and more than 30 Air Force personnel were treated for riot-related injuries.[90] Serious racial clashes also beset the Marine Corps—in July 1969 at Camp Lejeune, North Carolina, and in August 1969 at Kaneohe Naval Air Station, Honolulu.[91]

While some social scientists concluded that the individual performance of black combat soldiers in Vietnam was on a par with that of white soldiers, it was also obvious that intergroup relations reflected some of the same tension that had developed in American society.[92] Impressions about race relations in Vietnam are largely anecdotal, since intergroup relations during that era were not subjected to the rigorous scrutiny that social scientists had applied to the World War II and Korean experiences. Accounts were often conflicting. According to one group of investigators:

Racial tension, like drugs, was a problem the armed forces inherited from civilian society. The authoritarian nature of military society, which placed a white and heavily southern command structure over a young and substantially

87. Ibid., pp. 129–58.
88. *New York Times,* September 4, 1968.
89. Two of the most serious incidents occurred in the same week in October 1972. A fifteen-hour melee between black and white seamen on the carrier *Kitty Hawk* was followed by racial turmoil on the carrier *Constellation.* Within a month racial incidents took place on board an assault ship (*Sumpter*) and an oiler (*Hassayampa*). By an unofficial estimate, 196 men, mostly blacks, were arrested. Robert W. Mullen, *Blacks in America's Wars: The Shift in Attitudes from the Revolutionary War to Vietnam* (Monad Press, 1973), pp. 83–84.
90. An account of this riot and an excellent overall treatment of the integration of the Air Force are included in Gropman, *The Air Force Integrates,* pp. 215–16.
91. Adam Yarmolinsky, *The Military Establishment: Impacts on American Society* (Harper and Row, 1971), p. 344.
92. For example, Moskos concluded: "My observations as well as those of others found no differences in white or black combat performance in Vietnam"; *The American Enlisted Man,* p. 130. Butler and Wilson reported: "The Vietnam war and the era of the draft saw the development of racial polarization in an integrated military during the late 1960s"; John S. Butler and Kenneth L. Wilson, "The American Soldier Revisited: Race Relations and the Military," *Social Science Quarterly,* December 1978, p. 452.

black enlisted population, aggravated racial hostilities and hindered official efforts to overcome discrimination. . . .

Blacks found community not in the service, but among themselves. The intense racial consciousness of young blacks was hard for white officers to understand. "What defeats me," said a battalion commander, "is the attitude among the blacks that 'black is right' no matter who is right or wrong." At an official "rap" session in Germany, a general was told that all whites were pigs and he, being white, was also a pig. "I burned buildings in Chicago and shot whitey, and it doesn't bother me one bit," added the black soldier. "I'd just as soon shoot at whitey as the VC."

Black consciousness fed white racism. One soldier recalled hearing blacks called "reindeer," "Mau Mau," "jig," "spook," "brownie," "warrior," "coon," "spade," and "nigger." There was an atmosphere of mutual fear and distrust. As a white Green Beret recalled, "Blacks pretty much stuck to themselves and hated everyone else. I turned into a bigot in the military when I was on a bus with mostly black GIs, and they harassed and accosted me. You were told not to walk around the barracks at night, especially alone." A black GI responded that "whites think every time colored guys get together, well, he's a Panther, he's a militant."

Racial conflict was especially troublesome in Vietnam. By 1970, black unrest had begun to hinder the fighting effort. There was fear among white officers that black soldiers would turn their guns around and, as the soldier had said, "shoot at whitey" instead of the Viet Cong. Some, in fact, did. In one incident, two white majors were shot trying to get some black soldiers to turn down a noisy tape recorder.[93]

But others told a somewhat different story. Reflecting on a visit with military units in Vietnam, a journalist wrote:

There are reports of racial discrimination, racial fights and instances of self-segregation, but most Negroes interviewed said these instances were greatly outweighed by racial cooperation. . . .

It is in the front-lines that commonly-shared adversity has always sprouted quickly into group loyalty and brotherhood. And whether between white and white, Negro and Negro, or Negro and white, Vietnam is no exception to the tradition of battlefield brotherhood. . . .

And a long-time front-line observer said: "It's the most natural thing in the world to come out closer than brothers after a few days on the line. . . . This will make any two people brothers."[94]

OF ALL THE SOCIAL FORCES operating in the 1960s, it was the collision of the civil rights movement, the antiwar movement, the War on Poverty, federal legislation to create a "balanced society," and the "channeling" policies of the Selective Service System that aroused public awareness

93. Baskir and Strauss, *Chance and Circumstance*, pp. 137–38.
94. *New York Times*, April 29, 1968.

of "equality of service" and group participation within the military. This awareness soon became concern on the part of some about the unfairness of a disproportionately black military and uneasiness on the part of others about both the capability and the stability of a racially unbalanced fighting force. Some black leaders preferred to think that white fears of black overrepresentation in the military were due to the inherent racism of white America. The apprehension of many white Americans in the late 1960s was an emotional reaction; it was a part of the times, provoked by the threat of racial violence that surfaced in many cities each summer. And it was one of the major obstacles that the architects and proponents of an all-volunteer military would have to clear away.

CHAPTER THREE

BLACKS IN THE POST-VIETNAM ARMY

IN MARCH 1969, after close to a year of study, President Nixon's Commission on an All-Volunteer Force (the Gates Commission) issued its conclusion:

We unanimously believe that the nation's interests will be better served by an all-volunteer force. . . . We have satisfied ourselves that a volunteer force will not jeopardize national security, and we believe it will have a beneficial effect on the military as well as the rest of our society.[1]

Following a period of intense and often emotional debate, Congress agreed to remove the president's induction authority effective July 1, 1973, setting in motion a bold experiment—maintaining superpower-size military forces by voluntary means is a monumental task, unprecedented in any country's history.

Recruitment and Retention

Candidate Richard Nixon, in first proposing an end to the draft during the 1968 presidential campaign, characterized as "sheer fantasy" the notion that "a volunteer army would be a black army, so it is a scheme to use Negroes to defend a white America."[2] Some Americans disagreed and wondered whether the Selective Service System would be supplanted by a class-based structure that freed the white majority from their democratic duties. Some supported the all-volunteer concept primarily on the principle of individual freedom from servitude. Still others undoubtedly foresaw the approaching demise of protective draft

---

1. *The Report of the President's Commission on an All-Volunteer Armed Force* (Macmillan, 1970), p. i.
2. Richard M. Nixon, "The All-Volunteer Armed Force," address given over the CBS radio network, October 17, 1968, quoted in Gerald Leinwand, ed., *The Draft* (Pocket Books, 1979), p. 106.

deferments and found the proposed switch to a draft-free system especially appealing. With the end of the war in Vietnam, many arguments that the "volunteer experiment" would be unfair faded quickly, along with apprehension about a predominantly "disadvantaged" fighting force.

During the phase-out of compulsory service, the relative number of black volunteers increased steadily, though slowly. After just two years without conscription, however, it was obvious that the composition of the military's rank and file was undergoing a transformation: far fewer college-trained men were signing up, proportionately fewer recruits had above-average and below-average aptitude test scores, and there were indications that the socioeconomic character of the force was changing. Even more striking was the sudden leap in the proportion of blacks joining the Army. Over one out of four new soldiers was black—more than double the percentage of black Army recruits in 1970, the year the Gates Commission confidently predicted that "the composition of the military will not be fundamentally changed by ending conscription."[3] And all signs now pointed to a substantial increase in the proportion of blacks in uniform.

The situation was described in a Congressional Research Service publication:

DoD has repeatedly stated that it is not concerned with the racial breakdown of the Armed Forces and regards any action taken to limit enlistments by race as a violation of the concept that each individual must be measured on his own worth regardless of color. Congress, however, continues to be concerned that the Armed Forces may be becoming disproportionately composed of individuals who have lower socioeconomic status or who are members of racial/ethnic minorities.[4]

The Defense Manpower Commission was created by Congress in 1973 and directed to conduct a comprehensive study of the overall manpower requirements of the Department of Defense, including "the implications for the ability of the armed forces to fulfill their mission as a result of the change in the socioeconomic composition of military enlistees since the enactment of new recruiting policies provided for in Public Law 92-129 and the implications for national policies of this change in the composition

---

3. *Report of the President's Commission on an All-Volunteer Armed Force*, p. i.
4. Robert L. Goldich, "All-Volunteer Military Force," Issue Brief Number IB73021 (Washington, D.C.: Congressional Research Service, 1973), p. 4.

of the armed forces."[5] The Department of Defense was likewise directed by Congress to submit annual reports on "population representation in the All-Volunteer Force"—that is, the geographic, economic, educational, and racial composition of new recruits and members of the active force—at the end of each fiscal year.[6]

In recent years the representation issue has come to be associated primarily with the overrepresentation of blacks in the Army—this is probably the major concern among all those expressed about "military representation."[7] Of course, black participation is one of the oldest, most enduring military manpower "problems" but the recent interest is perhaps best explained by the coincidence of the steadily rising trend of black enlistments with the removal of Selective Service controls—for the first time in the history of an integrated force—on the social composition of the armed forces.

As table 3-1 shows, all the services entered the 1980s with a greatly increased proportion of black enlisted personnel and officers. In the Army, the proportion of blacks increased every year in the 1970s. At the

---

5. Defense Manpower Commission, *Defense Manpower: The Keystone of National Security,* Report to the President and Congress (Government Printing Office, 1976), p. 156.

6. See "Population Representation in the All-Volunteer Force" (Office of the Assistant Secretary of Defense for Manpower, Reserve Affairs, and Logistics, 1974 to present).

7. Kenneth J. Coffey and Frederick J. Reeg, "Representational Policy in the U.S. Armed Forces," in *Defense Manpower Commission Staff Studies and Supporting Papers,* vol. 3, p. D-12. See also Morris Janowitz and Charles C. Moskos, Jr., "Racial Composition in the All-Volunteer Force," *Armed Forces and Society,* vol. 1 (November 1974), pp. 109–23; Alvin J. Schexnider and John Sibley Butler, "Race and the All-Volunteer System: A Reply to Janowitz and Moskos," *Armed Forces and Society,* vol. 2 (Spring 1976), pp. 421–32; Mark J. Eitelberg, *Evaluation of Army Representation,* TR-77-A-9 (Alexandria, Va.: U.S. Army Research Institute for the Behavioral and Social Sciences, 1977); Mark J. Eitelberg, "American Youth and Military Representation: In Search of the Perfect Portrait," *Youth and Society,* vol. 10 (September 1978), pp. 5–31; Alvin J. Schexnider, "The Black Experience in the American Military," *Armed Forces and Society,* vol. 4 (Winter 1978), pp. 329–34; and numerous other references in the popular media, academic journals, research monographs, and government reports.

The General Accounting Office, in fact, recently criticized the Department of Defense for not making information on minorities (and females) available to Congress and the public on a regular basis. "Because of the increased numbers and proportions of minorities and females in the Armed Forces and the possible impact of these changes on manpower effectiveness, the Congress should be provided more information on this issue," the GAO reported. See General Accounting Office, "Minority and Female Distribution Patterns in the Military Services," FPCD-81-6 (GAO, Federal Personnel and Compensation Division, December 18, 1980).

Table 3-1. Blacks as a Percentage of the Armed Forces, by Service, Selected Fiscal Years, 1942–81[a]

Fiscal year	Army			Navy			Marine Corps			Air Force			All services		
	Enlisted	Officer	Total	Enlisted	Officer	Total	Enlisted	Officer	Total	Enlisted	Officer	Total	Enlisted	Officer	Total
1942[b]	6.2	0.3	5.8	n.a.	n.a.	n.a.	n.a.	n.a.	n.a.	[b]	[b]	[b]	n.a.	n.a.	n.a.
1945[b]	9.3	0.8	8.4	4.8	0.0	4.0	n.a.	n.a.	n.a.	[b]	[b]	[b]	n.a.	n.a.	n.a.
1949	11.1	1.9	10.1	4.4	*	5.3	2.5	*	2.3	6.1	0.6	5.3	7.5	0.9	6.7
1964	11.8	3.3	10.9	5.9	0.3	5.3	8.7	0.3	7.9	10.0	1.5	8.6	9.7	1.8	8.7
1968	12.6	3.3	11.5	5.0	0.4	4.5	11.5	0.9	10.7	10.2	1.8	8.9	10.2	2.1	9.2
1970	13.5	3.4	12.1	5.4	0.7	4.8	11.2	1.3	10.2	11.7	1.7	10.0	11.0	2.2	9.8
1971	14.3	3.6	12.9	5.4	0.7	4.8	11.4	1.3	10.4	12.3	1.7	10.5	11.4	2.3	10.2
1972	17.0	3.9	15.0	6.4	0.9	5.7	13.7	1.5	12.5	12.6	1.7	10.8	12.6	2.3	11.1
1973	18.4	4.0	16.3	7.7	1.1	6.8	16.9	1.9	15.4	13.4	2.0	11.5	14.0	2.5	12.4
1974	21.3	4.5	19.0	8.4	1.3	7.5	18.1	2.4	16.5	14.2	2.2	12.1	15.7	2.8	13.9
1975	22.2	4.8	19.9	8.0	1.4	7.2	18.1	3.0	16.7	14.6	2.5	12.5	16.1	3.1	14.3
1976	24.3	5.3	21.9	8.1	1.6	7.3	17.0	3.5	15.6	14.7	2.8	12.7	16.9	3.5	15.1
1977	26.4	6.1	23.9	8.7	1.9	7.9	17.6	3.6	16.2	14.7	3.2	12.7	17.9	4.0	16.0
1978	29.2	6.4	26.3	9.4	2.2	8.5	19.0	3.7	17.6	14.9	3.6	13.0	19.3	4.3	17.3
1979	32.2	6.8	28.9	10.7	2.3	9.7	21.5	3.9	19.8	15.8	4.3	13.8	21.2	4.7	19.0
1980	32.9	7.1	29.6	11.5	2.5	10.4	22.4	3.9	20.6	16.2	4.6	14.1	21.9	5.0	19.6
1981	33.2	7.8	29.8	12.0	2.7	10.8	22.0	4.0	20.2	16.5	4.8	14.4	22.1	5.3	19.8

Sources: Data for 1942 and 1945 from Ulysses Lee, Jr., *The United States Army in World War II*, Special Studies: *The Employment of Negro Troops* (Office of the Chief of Military History, U.S. Army, 1966), p. 415. Data for 1949–70 from Department of Defense, *The Negro in the Armed Forces: A Statistical Fact Book* (Office of the Deputy Assistant Secretary of Defense for Equal Opportunity, 1971). Data for 1971–80 provided by Department of Defense, Defense Manpower Data Center.

n.a. Not available.

* Less than 0.05 percent.

a. Percentage computations are based on the total active force in June for 1942, 1945, 1949, and 1971–75; in December for 1964, 1968, and 1970; and in September for 1976–81. Officers include commissioned and warrant officers.

b. Army computations for 1942 and 1945 include Air Force personnel.

end of fiscal 1981, one out of every three soldiers was black—three times the percentage of black enlisted personnel in 1949, when the Fahy Committee first urged the Army to remove its racial quota. Blacks constituted lower proportions in the other services; still, close to 20 percent of all those on active duty were black, a proportion substantially greater than the architects of the all-volunteer force had envisioned.

Renewed interest in the "black problem" was also stirred by the concurrent changes in the reserves. The Gates Commission "recognized from its first meeting the need for special attention to the problem of the reserve forces."[8] However, throughout the Vietnam era the reserves—used to supplement the active-duty forces, help maintain domestic peace, and assist in time of civil disaster—were a haven for white young men who wanted to avoid being drafted. The very small proportion of blacks serving in the reserves was a minor issue of the day, since it further demonstrated the social inequity of conscription.

The proportion of blacks in the selected reserve force increased in parallel fashion with changes in the active force, but in 1981 blacks were still *under*represented in two components, and the upward trend of black participation appeared to have slowed (see appendix table B-1). Yet the changes were, first, not anticipated, and second, in an opposite direction from the perceived "norm."

A closer look at the recent recruiting experiences of the Army (table 3-2) reveals that the proportion of black enlisted volunteers reached an all-time high of almost 37 percent in 1979, almost three times the proportion of blacks in the general youth population (about 13 percent).[9]

---

8. *Report of the President's Commission on an All-Volunteer Armed Force*, p. 95.

9. Statistical comparisons of military and civilian populations to determine representativeness are not always consistent. Conventional studies of population representation in the armed forces use the general population (segmented by race, age, and sex) as the standard or reference population. However, various groups can be used as the national civilian standard for comparison (for instance, the civilian labor force or divisions of it, the population that served during the draft, the general population of military-age youth, the general population, high school graduates), and various aggregations and combinations of groups from the armed forces can be used for proportional measurement, from the entire Department of Defense down to the smallest identifiable unit (for instance, total armed forces, separate services, recent accessions, total enlisted force, officers, males only, occupational specialties, broad skill groups, geographical distribution of personnel according to branch units and echelons, or the general distribution of group members by rank within units and subdivisions of units to the smallest group, an infantry platoon or squad). It has even been suggested that standards for comparison be drawn from the conscripted forces of earlier years, though this is not a truly representative configuration of the American people. Another case is often made for using 1964 as a base year or

In 1980 enlistments of blacks dropped somewhat, followed by a sharp decline in 1981. In fact, in absolute terms, fewer blacks entered the Army in fiscal 1981 than in any one-year period since the end of conscription and fewer black males than since the early 1960s.

The reasons for the sudden change in the black enlistment rate are not clear. The drop may be due to an increase in the supply of white youths resulting from worsening conditions in the job market, substantial increases in entry-level pay, and enlistment incentives that may favor whites (such as more generous educational benefits designed to attract "quality" male enlistees). On the other hand, the possibility that "special efforts" were made by Army recruiters to bring in white enlistees, in response to criticism about the racially unbalanced Army and the declining quality of the force (as measured by aptitude tests), should not be discounted. In 1981 the Army announced that it was shifting recruiters from urban areas to suburban locations near high school and college campuses in an effort to obtain "better" recruits.[10]

Once in the military, blacks are more inclined than whites to choose it as a career. When allowed, blacks have reenlisted at greater rates than their white counterparts throughout the recent recorded history of the armed forces.[11] In the Army, for instance, the reenlistment rates for both

---

benchmark for comparison, since it was both before the all-volunteer force and the last peacetime year before the war in Vietnam. Since officers tend to differ markedly from enlisted personnel (as the white-collar civilian labor force differs from the blue-collar force), the common practice of using only the enlisted force in comparisons with the general civilian population is sometimes criticized.

Should the percentage of blacks in the armed forces (recent accessions, total force, officers only, or total enlisted force), then, be compared with (1) the total proportion of blacks in the population (11 to 12 percent); (2) the proportion of blacks in the general population aged eighteen to twenty-three (13 to 14 percent); (3) the proportion of blacks among high school graduates between eighteen and twenty-two (12 percent); (4) the proportion of blacks in the military-age population who would be expected to qualify for enlistment (between 5 and 8 percent); (5) the proportion of blacks in the eighteen- to twenty-three-year-old noncollege male population (20 percent); or (6) the proportion of blacks in the manufacturing and construction trades (21 percent), the total blue-collar sector (14 percent), or some other area of the labor force? See Mark Jan Eitelberg, "Military Representation: The Theoretical and Practical Implications of Population Representation in the American Armed Forces" (Ph.D. dissertation, New York University, 1979), pp. 26–33, 86–98; Coffey and Reeg, "Representational Policy," p. D-20; and Richard V. L. Cooper, *Military Manpower and the All-Volunteer Force,* R-1450-ARPA (Santa Monica: Rand Corp., 1977), p. 205.

10. Larry Carney, "Recruiters Will Move Nearer Campuses," *Army Times,* September 7, 1981.

11. See, for example, "Retention Rates and Composition of the Male Enlisted Force

Table 3-2. Black Enlisted Entrants to the Army, by Sex, Selected Fiscal Years, 1954–81[a]

Fiscal year	Male		Female		Total	
	Number	Percent	Number	Percent	Number	Percent
1954[b]	34,617	9.9	n.a.	n.a.	n.a.	n.a.
1964[b]	30,534	14.0	n.a.	n.a.	n.a.	n.a.
1971	41,326	14.1	1,161	20.7	42,487	14.2
1972	26,599	15.1	1,055	17.7	27,654	15.2
1973	38,159	19.6	1,574	18.9	39,733	19.6
1974	46,250	28.0	2,987	19.8	49,237	27.4
1975	37,491	23.2	3,558	19.2	41,049	22.8
1976	40,710	25.0	2,810	17.8	43,520	24.3
1976[c]	14,619	30.1	955	22.4	15,574	29.5
1977	44,900	30.1	3,163	21.6	48,063	29.4
1978	36,624	34.9	5,239	30.3	41,863	34.2
1979	40,030	36.1	7,010	40.9	47,040	36.7
1980	37,790	28.0	8,775	39.6	46,565	29.7
1981	25,328	25.7	6,635	36.6	31,963	27.4

Sources: Data for 1954 and 1964 Army accessions from Bernard D. Karpinos, *Male Chargeable Accessions: Evaluation by Mental Categories (1953–1973)*, SR-ED-75-18 (Alexandria, Va.: Human Resources Research Organization, January 1977), pp. 33–35. All other years derived from data provided by the Defense Manpower Data Center.

n.a. Not available.

a. Enlisted entrants include inductees and enlistees without prior service.

b. Data for 1954 and 1964 include blacks and other (nonwhite) racial minorities who entered military service between January and December of each of those years.

c. Fiscal 1976 transition quarter (July through September).

first-term and career blacks (who were eligible to reenlist) far exceeded the comparable rates for whites each year after the end of conscription (see appendix table B-2). In fact, the proportion of blacks among all Army reenlistments doubled between 1972 and 1981, to a point where more than one out of every three was black.

Profile of Black Volunteers

Concurrent with the growth in the proportion of blacks in the armed forces has been an increase in criticism, mostly by whites, of the quality of volunteers. While the concept of quality is vague, concern centers on a perceived deterioration in the caliber of recruits as measured by standardized entry test scores, level of education, socioeconomic status,

by Race and Year of Entry to Active Service as of 30 June 1973," Manpower Research Note 73-13 (Office of the Deputy Assistant Secretary of Defense for Equal Opportunity, September 1971), pp. 174–229.

rates of attrition, and incidence of indiscipline. At the same time, many blacks continue to measure racial discrimination in the armed forces in terms of fewer "good" job opportunities, slower promotions, particularly in the enlisted force, and underrepresentation in the officer corps. As summed up by one observer: "The continuing stratification of the Army—blacks at the bottom, whites at the top, blacks on the firing line, whites manning technical posts—suggests that the real problem has to do with unkept promises rather than with the dangers of racial imbalance."[12]

Military Aptitude

Pencil-and-paper tests have been used by the armed forces since World War II to screen draft registrants and applicants for enlistment for whom the probability of success in the service is low. The tests are also used to determine whether recruits are eligible for specific job training. The military justifies the tests in statistical terms: "while aptitude tests are not perfect predictors, they do enhance the probability that the services will select the best people from the pool of applicants and will assign them to jobs in which they are likely to succeed."[13]

Historically, black recruits have not performed as well as whites on military aptitude tests. Table 3-3, for instance, depicts the distribution of white and nonwhite male enlisted entrants by Armed Forces Qualification Test (AFQT) category for selected years.[14] These historical data show that in the entire period under 9 percent of nonwhite male enlisted recruits placed in the above-average categories (I and II), compared to about 39 percent of the white males; at the other end of the scale, roughly one of every six whites and one of every two minorities attained scores that placed them in the below-average category IV.[15]

The sharp increase in the proportion of recruits, particularly non-

12. Joseph Kelley, "The Army Is Looking for a Few White Men," *In These Times*, February 27–March 4, 1980, p. 24.
13. Department of Defense, *Aptitude Testing of Recruits*, Report to the House Armed Services Committee (Office of the Assistant Secretary of Defense for Manpower, Reserve Affairs, and Logistics, 1980).
14. "Nonwhite" rather than black is used for this comparison because of data availability.
15. AFQT categories are percentile groupings, the boundaries of which have been selected for statistical convenience. Although AFQT category data are used by the military services mainly for historical comparisons, increasingly they have been used by outside observers as barometers of quality.

whites, who scored in the category IV range in the latter part of the 1970s caused much concern. As the table shows, the proportions of white recruits in that category, who are considered by the services to require more training and to present greater disciplinary problems, grew from about 8 percent in 1976 to over 22 percent in 1980, while the comparable numbers for nonwhites grew from roughly 23 percent to 58 percent.[16] This development, probably more than any other factor, has been cited by critics of the all-volunteer force as evidence that the concept has failed. Controversy has arisen, however, regarding the appropriateness of military aptitude tests in general and their racial implications in particular. This is discussed in more detail in chapter 5.

Level of Education

Although black recruits have not performed as well as their white counterparts on the pencil-and-paper tests, in recent years they have surpassed whites in formal education. Oddly enough, the latter has largely resulted from the former. Since by current standards high school dropouts must attain higher scores on entry tests, proportionately fewer black dropouts now qualify;[17] hence a larger proportion are high school

16. The Pentagon did not realize until 1980 what was happening. With the introduction of a new version of the standardized entry test in 1976, errors in converting raw test scores into percentile scores caused the latter to be overstated, with the result that many recruits who would otherwise have been ineligible were accepted by the military services. The magnitude of the errors was substantial; for example, in contrast to the original belief that only 5 percent of the recruits who entered the armed forces in fiscal 1979 had scored below the thirty-first percentile (category IV), the corrected scores placed 30 percent in that category. The Army was the most seriously affected—close to half its recruits scored in the below-average category IV rather than 9 percent, as reported originally. Department of Defense, "Aptitude Testing of Recruits," Report to the House Committee on Armed Services, July 1980, p. 10. Tabulations of Army recruitment data for fiscal 1979 also reveal substantial differences by race; 68 percent of the black male recruits actually scored in category IV rather than 20 percent, as reported originally. The comparable figures for nonblack males were 38 percent and 6 percent, respectively.

The military apparently does not have a monopoly on test calibration problems. By one account, the widely advertised decline in scholastic aptitude test (SAT) scores has been understated as a result of improperly equating differences in test forms. See Lyle V. Jones, "Achievement Test Scores in Mathematics and Science," *Science,* vol. 213 (July 24, 1981), p. 412.

17. Based on entry standards in effect in fiscal 1981, only 7 percent of all black high school dropouts (aged eighteen to twenty-three) would be likely to qualify for enlistment; in contrast, 69 percent of black high school graduates in that age group would meet eligibility standards. About 42 percent of all white high school dropouts and 96 percent of white high school graduates would probably qualify for enlistment in the Army. Qualification rates are discussed in chapter 5.

Table 3-3. Percentage Distribution of Male Enlisted Entrants to All Services, by AFQT Category and Race, Selected Years, 1953–81[a]

Year and race	AFQT category[b]				Total	Number (thousands)
	I	II	III	IV		
1953						
White	8.6	27.0	35.4	29.0	100.0	668.0
Other races	0.6	4.1	18.7	76.6	100.0	86.5
1958						
White	9.5	28.3	45.8	16.4	100.0	393.9
Other races	0.8	7.1	41.1	51.0	100.0	32.1
1964[c]						
White	6.8	35.6	47.9	9.7	100.0	371.6
Other races	0.4	7.9	52.6	39.1	100.0	45.3
1966[d]						
White	7.4	36.9	41.6	14.1	100.0	842.9
Other races	0.6	7.2	38.4	53.8	100.0	90.2
1968						
White	7.2	36.0	38.3	18.5	100.0	713.3
Other races	0.4	6.9	30.8	61.9	100.0	99.5
1973[e]						
White	3.9	35.8	53.9	6.4	100.0	256.4
Other races	0.4	12.0	65.0	22.6	100.0	68.9
1976						
White	4.1	35.2	52.3	8.4	100.0	292.0
Other races	0.6	13.0	63.4	23.0	100.0	70.7
1978						
White	4.7	32.9	43.7	18.7	100.0	197.2
Other races	0.4	9.0	38.9	51.7	100.0	71.5
1980						
White	3.8	29.4	44.1	22.7	100.0	227.4
Other races	0.3	7.5	34.2	58.0	100.0	79.4
1981						
White	3.7	36.0	48.4	11.9	100.0	219.3
Other races	0.5	11.7	46.8	41.0	100.0	63.4
1953–72						
White	7.3	32.5	43.1	17.1	100.0	9,720
Other races	0.5	6.5	38.6	54.4	100.0	1,215
1973–81						
White	3.9	33.4	49.4	13.3	100.0	2,302
Other races	0.4	10.4	50.6	38.6	100.0	699
1953–81						
White	6.7	32.7	44.2	16.4	100.0	12,022
Other races	0.5	7.9	42.9	48.7	100.0	1,914

graduates. This is most apparent in the Army and in the Navy, where the proportion of black recruits in fiscal 1981 with a high school diploma exceeded the comparable proportion of white recruits by a substantial margin (see appendix table B-3).

This trend has attracted some attention since it is well known that the educational levels of American blacks trail behind those of their white counterparts. While the gap has narrowed in recent years, blacks' educational attainment is still markedly lower than that of whites. "In point of fact," notes one observer, "today's Army enlisted ranks is the only major arena in American society where black educational levels surpass those of whites and by a significant degree."[18]

Socioeconomic Status

Whether the criticism that a force composed solely of volunteers will put an unfair burden on the nation's socially and economically disadvantaged is valid is difficult to determine, first, because there is little agreement on the definition of social and economic disadvantage, and second, because the armed services do not routinely collect information on even the obvious indicators. Nevertheless, survey results have verified the self-evident: by most measures, black volunteers are of a lower socioeconomic status than their white counterparts. This is confirmed by a comparison of estimated family income of a sample of volunteers who entered military service in 1979, which revealed that, of those responding, the family income of three out of four blacks and only

18. Charles C. Moskos, "How to Save the All-Volunteer Force," *The Public Interest*, no. 61 (Fall 1980), p. 77.

Notes for table 3-3.
Sources: Armed Forces Qualification Test (AFQT) category distributions for 1953–73 derived from data in Karpinos, *Male Chargeable Accessions*. All other distributions derived from data provided by Defense Manpower Data Center.

a. Percentage distributions for 1953 through 1973 include all male enlisted accessions (enlistees and inductees) without prior service who entered military service between January and December of the respective year or years. Percentage distributions for subsequent years cover the fiscal year. Draftees who failed the aptitude tests but who were declared administratively acceptable (on the basis of personal interviews and some additional testing) are included in AFQT category IV.

b. All applicants for enlistment are tested for aptitude. Test scores are used to classify applicants into one of five categories (I through V). Those in categories I and II are above average in aptitude; those in category III are average; those in category IV are below average, but still eligible for enlistment; and those in category V (not shown) are at the very bottom of the scale and not eligible for service. The AFQT category distributions in this table incorporate scores that were refigured to correct for calibration errors made by the Defense Department between 1976 and 1980.

c. 1964 was the last peacetime year before the war in Vietnam.

d. The greatest influx of new recruits during any one-year period since World War II occurred in 1966.

e. 1973 was the first year of the all-volunteer force; the last draft call was issued by the Selective Service System in December 1972.

Table 3-4. Selected Characteristics of Males in the Armed Forces and Employed Male Civilians Aged 18 to 21, by Race, 1979
Percent

	White						Black					
Characteristic	Full-time employed[a]	All services	Army	Navy	Marine Corps	Air Force	Full-time employed[a]	All services	Army	Navy	Marine Corps	Air Force
Education of parent												
Less than 12 years	21	20	27	18	14	17	48	30	36	27	36	13
12 years	53	46	46	44	52	49	37	45	42	66	51	33
13 years or more	26	34	28	39	34	35	15	25	22	7	13	54
Occupation of parent[b]												
Professional or managerial	23	26	23	31	20	29	13	14	17	*	21	8
Sales, clerical	13	12	8	13	20	10	7	8	10	35	*	*
Blue collar	53	51	58	45	54	53	52	54	48	41	58	84
Service	7	10	11	11	6	9	21	23	25	23	21	9
Educational expectations												
Less than 12 years	10	2	3	3	1	1	13	1	1	*	*	*
12 years	50	29	36	24	41	20	41	18	23	6	21	6
13–15 years	25	24	27	26	16	20	26	29	37	11	34	9
16 years or more	15	45	34	47	43	59	21	52	40	83	45	84
"Knowledge of the world of work" score[c]												
0–5	18	15	21	12	12	9	55	39	44	24	46	12
6	15	13	21	9	12	10	16	10	14	*	10	7
7	23	22	19	39	17	29	17	20	22	16	13	22
8–9	44	50	39	39	58	53	12	31	19	60	32	59

Sources: Derived from Choongsoo Kim and others, "The All-Volunteer Force: An Analysis of Youth Participation, Attrition, and Reenlistment," prepared for the Employment and Training Administration, Department of Labor (Ohio State University, Center for Human Resource Research, 1980), p. 19; and data provided by Defense Manpower Data Center, April 1981. Figures are rounded.
* Less than 0.5 percent.
a. Excludes those who still attend high school or are enrolled in college full time.
b. Excludes farming.
c. "Knowledge of the world of work" is a test in which respondents are asked to identify the kinds of work done in nine occupations. The scores indicate how many occupations were correctly identified.

one of every two whites was below $16,000 a year (see appendix table B-4). The same survey revealed that about 18 percent of black recruits and 7 percent of the whites reported that some of the family income was "from public assistance or welfare" sources.[19]

More surprising, however, are the data shown in table 3-4 indicating that blacks who enter the military services have "better" backgrounds and "better" credentials than blacks who are employed full time in the civilian labor force while the comparable white groups tend to be more similar.[20] For example, more black servicemen than civilians came from families where a parent had completed high school, proportionally fewer scored below the mean on a test of "knowledge of the world of work," and many more expected to complete college. White servicemen and full-time civilian workers, on the other hand, tended to be more alike in the distribution of their parents' occupation and education and in their knowledge of the world of work.[21] "The most striking finding on the participation of youth in the military," according to a group of scholars, "is the higher participation rate among minorities [blacks and Hispanics] from higher socioeconomic backgrounds and with better qualifications."[22]

A separate comparison reveals significant differences between the services. For example, table 3-4 shows that Army males, both black and white, tend to be from a lower social background and have lower qualifications than the armed forces norms, but with few exceptions they still hold an advantage over their respective civilian counterparts.

Attrition Rates

The first-term attrition rate—the percentage of enlisted personnel discharged before completing their first three years of service—has been

19. Derived from the 1979 Department of Defense Survey of Personnel Entering Military Service.
20. In this comparison those who are still in high school or attend college full time are excluded from the civilian work force.
21. "Knowledge of the world of work" is a test in which respondents are asked to identify the kinds of work done in nine occupations. The scores indicate how many occupations could be correctly identified. For a discussion of the test, see Herbert S. Parnes and Andrew I. Kohen, "Occupational Information and Labor Market Status: The Case of Young Men," *Journal of Human Resources,* vol. 10 (Winter 1975), pp. 44–55.
22. Choongsoo Kim and others, "The All-Volunteer Force: An Analysis of Youth Participation, Attrition, and Reenlistment," prepared for the Employment and Training Administration, Department of Labor (Ohio State University, Center for Human Resource Research, 1980), p. 12.

52   Blacks and the Military

Table 3-5. Trends in Disciplinary Incidents in the Armed Forces, by Race and Sex, Fiscal Years 1978–81
Number of persons per thousand average monthly strength

Type of disciplinary incident and sex	1978		1979		1980		1981	
	Black	White	Black	White	Black	White	Black	White
Unauthorized absence								
Male	42.6	33.3	40.4	32.3	46.0	34.9	39.4	31.5
Female	10.7	11.4	11.4	13.5	13.5	16.6	11.6	12.9
Designated deserter								
Male	13.5	17.4	14.4	18.0	15.6	18.2	12.4	15.5
Female	3.4	4.9	3.2	5.4	4.0	8.0	3.1	6.2

Source: Data provided by Office of the Assistant Secretary of Defense for Manpower, Reserve Affairs, and Logistics.

considered one of the most problematic aspects of the all-volunteer force since (1) the rate for males rose markedly in the mid-1970s (see appendix table B-5); (2) past efforts have failed to deal adequately with the problem; and (3) attrition means that to sustain force size more recruits are needed (at greatly increased costs).

Generally speaking, high school graduates are almost twice as likely as dropouts to complete the first three years of their enlistment periods. For those who entered the services in fiscal 1978, for example, the loss rates were almost identical for white and black male diploma holders and the attrition rates of black male dropouts were somewhat below those of whites. Overall, the proportion of black male volunteers who do not complete their first three years is lower than the proportion of whites who similarly separate. This trend coincides with the increase in the educational level of black volunteers.

Indiscipline and Punishment

A disproportionate share of disciplinary incidents and punitive actions occurs among racial minorities. Recent trends can be traced with the aid of table 3-5, which compares two measures of indiscipline: absenteeism and desertion.[23] The data show consistent differences: (1) males are far more apt than females to be AWOL or desert; (2) relatively more black

23. Absentees are those who are absent without leave for less than thirty days. Those who have been AWOL for thirty or more consecutive days are administratively classified as deserters. Only after the accused is convicted of the charge of desertion can the term "deserter" be applied in the strict legal sense.

Table 3-6. **Crime Rates of Army Personnel, by Race, Fiscal Years 1978–80**
Rates per thousand

Category	1978		1979		1980	
	White	Black	White	Black	White	Black
Crimes of violence[a]	2.8	12.7	2.8	11.7	3.1	11.8
Crimes against property[b]	10.1	17.6	9.9	18.9	12.8	22.1
Drug offenses[c]	36.6	54.6	34.7	56.7	35.9	54.2

Sources: *Equal Opportunity: Third Annual Assessment of Programs* (Office of the Deputy Chief of Staff for Personnel, Department of the Army, 1979), p. 56; and *Equal Opportunity: Fourth Annual Assessment of Military Programs* (ODCSP, 1980), p. 47. Data for 1980 provided by Department of the Army, Office of Equal Opportunity Programs. Figures are rounded.
a. Includes murder, rape, aggravated assault, and robbery.
b. Includes burglary, larceny, auto theft, and housebreaking.
c. Includes use, possession, sale, and trafficking.

males go AWOL but relatively fewer desert than their white counterparts; and (3) black females have the lowest incidence in both categories of indiscipline.[24] It is unclear both why black males are more inclined to take shorter unauthorized absences and white males to stay away for longer periods and why black females cause fewer problems than white females.

Distinctions by race are much more sharply drawn when comparing serious crime statistics for Army enlisted personnel. In that service blacks tend to commit more serious crimes by a substantial margin (see table 3-6) and, accordingly, to be overrepresented in Army prison populations:[25]

	Blacks as percent of	
Fiscal year	Enlisted personnel	Prison population
1977	26.4	51.0
1978	29.2	51.3
1979	32.2	51.2

The Army points out that even though blacks are overrepresented in the prison population the degree of overrepresentation is less than that

24. These results are fairly consistent across all the services, as appendix table B-6 shows.
25. *Equal Opportunity: Third Annual Assessment of Programs* (Office of the Deputy Chief of Staff for Personnel, Department of the Army, 1979), p. 56; and *Equal Opportunity: Fourth Annual Assessment of Military Programs* (ODCSP, 1980), p. 48. Moreover, the overrepresentation of blacks in the Army's prisons is a continuing trend; in 1972, for example, 17 percent of the Army was black, but 37 percent of the confined population was black; in 1976, 25 percent of the Army was black, but 50 percent of those confined were black.

in the civilian population. In 1979 the proportion of black Army prisoners was 1.6 times the proportion of blacks in the Army; on the other hand, the proportion of blacks in the Federal Bureau of Prisons systems was almost four times the proportion of blacks in the national population.[26]

It has been suggested that the overrepresentation of blacks in the Army's prison system is indirectly related to other disparities in black representation. The Southern Christian Leadership Conference in 1978 laid the blame for black overrepresentation in Army penal facilities on inequities in the criminal justice system—specifically the unrepresentatively low percentage of black officers (6.1 percent) and the predominance of prejudiced white officers from the South.[27] Officers make the initial decisions to deal with problems through minor punishment, court-martial, or early discharge. Administrative discretion thus plays a large part in the initial corrective action—and the decisions are mostly made by white officers. For example, blacks are greatly underrepresented not only in the officer corps, but throughout the entire justice system. In 1978 only one of the Army's forty-six trial court judges was black (and one female), only 4 percent of the Army's lawyers were black, and only 13 percent of the Army's military police were black.[28]

Blacks are overrepresented in the Army's confinement facilities, according to one study, not because the justice system itself is discriminatory but because of "cultural differences in the implementers of the systems and particular racial minorities." The study group concluded that "despite a decade of Army programs aimed at overcoming these cultural impediments to equality of opportunity and treatment, the objectives of those programs have not yet been achieved."[29]

The chief judge of the Court of Military Appeals echoed the sentiment: The court-martial rate for blacks might be reduced if there were more black judge advocates who could provide the military justice system with a better understanding of alleged offenses which are less a result of misconduct than of lack of knowledge or misperception of cultural differences.[30]

In some ways, the crime and confinement statistics are somewhat

26. *Equal Opportunity: Fourth Annual Assessment of Military Programs*, p. 48.
27. See Bill Drummond, "Army Concerned About Blacks' High Rates of Criminality," *Washington Post*, November 19, 1978.
28. Ibid.
29. Peter G. Nordlie and others, *A Study of Racial Factors in the Army's Justice and Discharge Systems*, HSR-RR-79/18-Hr, prepared for the DAPE-HRR, Department of the Army (McLean, Va.: Human Sciences Research, Inc., 1979), vol. 1, p. iv.
30. Remarks attributed to Chief Judge Robinson O. Everett, Jr., *Navy Times*, September 14, 1981.

surprising since the armed forces are apparently applying tighter enlistment screens to blacks than to whites. This is suggested by the relative proportions of recruits who are granted waivers on moral grounds; in fiscal 1981 white recruits were twice as likely to have entered with a moral waiver as their black counterparts (see appendix table B-7).

Of all enlisted personnel who were discharged in 1980 a smaller proportion of blacks were separated under "honorable" conditions, but the differences are very small (appendix table B-8). An examination of the causes of separation (appendix table B-9) shows that both black men and black women experienced lower total separation rates than their white peers. Black females, in fact, had lower separation rates than white females for all possible causes. The rates of black males exceeded those of white males in a few categories—notably, "marginal performance"—but the adjusted total separation rate for white males was still slightly higher.

Occupational Mix

Because blacks perform relatively poorly on the military's entry tests and because the aptitude testing system is used to match individuals with jobs, a disproportionate number of blacks have always served in the so-called soft, nontechnical jobs where training is minimal and advancement is often slow.[31]

Since the Vietnam War casualty controversy first erupted in the mid-1960s, the armed forces (especially the Army) have attempted to meet affirmative action goals for a more representative distribution among major occupational groupings. Most efforts have been concentrated on reducing the number of blacks who serve in the combat arms specialties—that is, the military jobs that are likely to bear the burden of casualties in wartime. The Army, as seen in table 3-7, has been successful in reducing the relative proportion of blacks assigned to those jobs. In fiscal 1981, 32.2 percent of the enlisted men in the Army were black, and 31.5 percent of the enlisted men assigned to combat occupations were black.

31. In addition to aptitude testing, other institutional policies and procedures in recruiting, training, and upgrading personnel influence the placement and utilization of the military's minority population. An analysis of this problem in the Navy and Marine Corps—and related issues involving the implementation of effective affirmative action and equal opportunity programs—can be found in Herbert R. Northrup and others, *Black and Other Minority Participation in the All-Volunteer Navy and Marine Corps* (Wharton School, University of Pennsylvania, 1979).

Table 3-7. Blacks as a Percentage of Male Enlisted Personnel Assigned to Major Occupational Areas in the Army and All Services, Selected Years, 1964-81[a]

Occupational category	1964[b]		1972		1976		1981	
	Army	All services	Army	All services	Army	All services	Army	All services
Infantry, gun crews, and seamanship specialists	19.3	16.4	19.0	17.5	23.9	22.4	31.5	27.5
Electronic equipment repairmen	11.2	5.6	12.2	5.2	16.8	6.9	25.5	10.2
Communications and intelligence specialists	9.2	6.7	11.6	7.5	23.8	15.5	31.3	22.8
Medical and dental specialists	16.6	12.0	16.7	11.6	21.6	15.4	32.6	23.9
Other technical and allied specialists	9.3	7.1	10.4	8.2	13.5	12.4	24.1	17.6
Administrative specialists and clerks	11.7	10.4	18.3	14.8	31.7	22.4	42.2	30.1
Electrical/mechanical repairmen	11.3	7.2	14.3	8.9	20.4	12.1	29.6	17.0
Craftsmen	11.2	9.7	14.5	11.8	16.0	12.1	25.9	15.7
Service and supply handlers	17.1	17.1	22.2	20.6	24.2	20.9	34.2	28.0
Nonoccupational and miscellaneous[c]	5.0	6.2	16.2	15.4	28.4	19.0	24.6	17.1
Blacks as percent of all male enlisted personnel	11.8	9.7	17.0	12.6	24.4	16.8	32.2	21.6

Sources: Data for 1964 from Department of Defense, *The Negro in the Armed Forces: A Statistical Fact Book*. All other distributions derived from data provided by the Defense Manpower Data Center.

a. Percentage distributions are based on enlisted force compositions as of December 1964, June 1972, and September 1976 and 1981.
b. Data for 1964 include both males and females.
c. "Nonoccupational" includes patients, prisoners, officer candidates and students, persons serving in undesignated or special occupations, and persons not yet occupationally qualified (service members who are in basic or occupational training).

But the proportion of blacks in all the services assigned to the combat category (27.5 percent) still exceeded the overall percentage of black enlisted men (21.6) by a considerable margin.[32] Percentages of blacks in each occupational category in the Navy, the Marine Corps, and the Air Force are given in appendix table B-10.

There is a definite pattern of black participation in major occupational areas. Blacks are overrepresented in administrative and clerical jobs and in the relatively unskilled service and supply handler categories. This is nothing new; throughout World War I blacks were assigned almost exclusively to service and supply jobs, and at the end of World War II blacks in the Army were still found predominantly in quartermaster, transportation, and engineer occupations.[33]

A closer examination of the twenty most common occupational subgroups in the Army reveals more about recent trends in black participation. In fiscal 1981 over half of all Army enlisted men assigned to supply administration (55.4 percent) and linemen (57.8 percent) occupational subgroups were black (appendix table B-11). Although blacks were slightly underrepresented in the infantry, 45 percent of all men assigned to artillery and gunnery skills (the third most common occupational subgroup) were black. Blacks also accounted for over 40 percent of army enlisted personnel in several "soft skills": unit supply, food service, administration, and personnel. On the other hand, less than 16 percent of the law enforcement (military police) subgroup was black, and blacks were noticeably underrepresented in armor and amphibious, combat operations control, combat engineering, track vehicle repair, and aircraft jobs.

When the occupational subgroups are subdivided into primary military occupational specialties, even wider discrepancies in representation are found (see appendix table B-12). For example, about half of all enlisted males in the cannon and missile crewmen specialties are black, and in several specialties under the personnel subgroup, the proportion exceeds 50 percent. In "functional support and administration" jobs, percentages range from a high of 61 (equipment records and parts) to a low of 16 (programmer/analyst). In other subgroups, too, the proportion of blacks

---

32. This is due principally to the concentration of blacks in the Marine Corps combat category. See appendix table B-10.

33. H. S. Milton, ed., *The Utilization of Negro Manpower in the Army*, Report ORO-R-11 (Chevy Chase, Md.: Operations Research Office, Johns Hopkins University, 1955), p. 563.

58 *Blacks and the Military*

within certain occupational specialties is considerably above or below that expected on the basis of the overall proportion of blacks in the enlisted force.

Advancement

Table 3-8 shows that there is a greater concentration of blacks in the lower ranks. This is partly because promotion is time-related and the large influx of blacks is a relatively recent phenomenon. But there is also evidence that at least between 1971 and 1975 white soldiers were promoted more rapidly than blacks. Of the Army enlisted personnel promoted in 1975, for example, whites were promoted to grade E-5 an average of 2.4 months sooner than blacks; to grade E-6, the difference was 3.0 months; to grade E-7, 6.5 months; to grade E-8, 14.2 months; and to grade E-9, 17.5 months.[34] This situation apparently persisted into the late 1970s, as evidenced by the adverse trend in promotion to grade E-7 (sergeant, first class), shown below as percentages of those eligible who were selected for promotion in fiscal years 1976–81:[35]

Race	1976	1977	1978	1979	1980	1981
White	38.1	47.6	36.7	33.1	34.2	28.6
Black	34.9	43.4	28.8	25.7	31.2	26.3
Other	44.9	48.0	37.3	30.9	35.1	28.0

According to the fiscal 1979 promotion board, "There was a smaller percentage of Blacks recommended as best qualified than those recommended from the other two racial categories."[36] The elements given the most weight in the promotion process include evaluation reports, job proficiency as measured by primary occupational specialty (PMOSE) and skill qualification test (SQT) scores, education, and "commendatory/derogatory" information. While no significant differences were found in evaluation reports of black and white soldiers in the grade of E-6, blacks

---

34. Department of the Army, "Measuring Changes in Institutional Racial Discrimination in the Army," Pamphlet 600-43, April 1977, p. C-20. Interestingly, the relationship between standardized entry test scores and speed of promotion for whites was reversed for blacks. High-scoring whites tended to be promoted more rapidly than low-scoring whites; low-scoring blacks, however, tended to advance faster than high-scoring blacks. Ibid., p. 4-5.

35. Data for 1976–79 from *Equal Opportunity: Fourth Annual Assessment of Military Programs*, app. 20. Data for 1980 and 1981 from Department of the Army.

36. *Equal Opportunity: Fourth Annual Assessment of Military Programs*, p. 32.

Table 3-8. **Blacks as a Percentage of Total Enlisted Personnel, by Pay Grade and Service, September 1981**

Pay grade	Army	Navy	Marine Corps	Air Force	All services
E-9	22.4	6.0	15.4	10.1	13.0
E-8	27.3	5.6	15.0	12.7	16.8
E-7	24.5	5.3	16.1	14.2	16.1
E-6	26.1	6.7	19.6	15.4	16.9
E-5	35.8	10.0	21.9	19.2	23.2
E-4	39.8	14.4	21.5	17.3	26.5
E-3	33.6	15.6	25.7	16.3	23.2
E-2	29.2	15.2	22.2	14.1	20.8
E-1	29.5	14.0	19.7	13.9	21.1
All pay grades	33.2	12.0	22.0	16.5	22.1

Source: Derived from data provided by Defense Manpower Data Center.

scored, on the average, about 7 points (6 percent) lower on the PMOSE and about 7 points (10 percent) lower on the SQT, about 7 percent fewer had any education beyond the high school level, and as a result of higher rates of indiscipline, there was "a higher density of derogatory information" in their files.[37] As the data indicate, the gap again started to narrow in the following year and was all but eliminated by 1981.

Underrepresentation in the Officer Corps

Blacks were virtually excluded from the officer corps until the end of World War II. As shown in table 3-1, in September 1981 blacks still constituted a relatively small percentage of officers in all four services—certainly, a percentage that in no way reflected the changes in racial composition that occurred in the enlisted ranks. Here again, some occupations do not offer the same career progression opportunities as others, and as table 3-9 shows, black officers tend to be concentrated in jobs (such as supply, procurement, and administration) not considered part of the mainstream and to be clustered in the lower ranks (see table 3-10).

Black underrepresentation in the officer corps is therefore an issue of concern to many people, yet it generally receives much less attention than black overrepresentation in enlisted personnel. This may be, first, because the proportion of black officers in the armed forces is roughly

37. Ibid., pp. 32, 33, 34.

Table 3-9. Blacks as a Percentage of Officers Assigned to Major Occupational Areas, by Service and Sex, September 1981[a]

Occupational category[b]	Army		Navy		Marine Corps		Air Force		All services	
	Male	Female	Male	Female	Male	Female	Male	Female	Male	Female
General officers and executives	5.0	50.0	1.1	6.0	0.2	0.0	1.9	8.3	1.5	6.6
Tactical operations officers	5.2	0.0	2.4	2.5	2.6	0.0	2.6	10.4	3.4	7.9
Intelligence officers	4.4	7.1	1.6	3.2	3.0	0.0	3.4	7.0	3.5	6.3
Engineering and maintenance officers	9.3	19.4	2.6	3.0	7.0	0.0	4.8	10.0	5.7	13.0
Scientists and professionals	6.0	5.5	2.6	4.7	5.0	0.0	4.3	9.8	4.2	6.5
Medical officers	5.3	10.2	2.4	4.2	c	c	3.4	7.9	4.0	7.9
Administrators	9.4	15.8	2.5	7.1	7.6	3.1	8.2	15.4	6.8	12.7
Supply, procurement, and allied officers	11.8	13.9	3.6	5.4	8.7	14.1	7.9	17.8	7.7	15.1
Nonoccupational[d]	3.8	10.3	0.0	0.0	3.4	5.9	2.6	1.9	3.0	4.5
Black officers as percent of all officers	7.2	13.8	2.6	4.8	4.0	4.6	4.2	10.6	4.9	10.3

Source: Derived from data provided by Defense Manpower Data Center.

a. As of September 1981, 8.2 percent of all officers were women. However, there is great variation in the relative proportion of female officers within the major occupational groups. For example, women account for 2.7 percent of all general officers and executives, and less than 1 percent of tactical operations officers. On the other hand, one out of eight officers assigned to administration and almost one out of three medical officers are female. Percentages include commissioned and warrant officers.

b. Approximately 25 percent of all Army officers and 12 percent of all Navy officers could not be identified by major occupational category.

c. The Navy provides the Marine Corps with medical support.

d. Includes patients, students, and others (such as those with duties unassigned, ROTC officers waiting to be placed on active duty, special assignment officers).

Table 3-10. Blacks as a Percentage of Total Officers, by Pay Grade and Service, September 1981

Type of officer and pay grade	Army	Navy	Marine Corps	Air Force	All services
Commissioned officers	8.1	2.6	3.7	4.8	5.3
O-7 and above	5.8	0.8	1.5	3.2	3.5
O-6	4.8	0.9	0.0	1.8	2.5
O-5	4.8	0.6	0.7	2.3	2.7
O-4	4.6	1.4	1.8	2.3	2.8
O-3	8.8	3.6	4.7	5.1	6.1
O-2	13.5	3.6	5.2	8.4	8.4
O-1	9.6	3.5	4.1	7.1	6.7
Warrant officers	5.9	5.1	7.5	...	5.9
All officers	7.8	2.7	4.0	4.8	5.3

Source: Derived from data provided by Defense Manpower Data Center.

in line with the proportion of all college graduates in the relevant age group who are black,[38] and second, because the proportion of blacks in the officer corps has been steadily rising (in the direction of representation).

The Army also argues that affirmative action efforts designed to increase the number of minority officers are only beginning to pay dividends. The recruitment of qualified minorities has been difficult, states the Army, largely because of intense civilian competition for minority college graduates and recruiting efforts by universities to enroll minorities who might otherwise enter precommission programs.[39] Another factor that could be contributing to black college graduates' lack of interest in the armed forces is the finding that as a group—often referred to as the "crossover generation"—they tend to be more suspicious of whites and to see more discrimination in the struggle for advancement than other members of the black community.[40] The fact that a military career would provide greater earnings than the average black college graduate could expect in the private sector suggests that economic considerations are subordinate to other less tangible factors.[41]

38. In 1979, 5.3 percent of all male college graduates aged twenty to twenty-nine were black. Bureau of the Census, *Current Population Reports,* series P-20, no. 356, "Educational Attainment in the United States: March 1979 and 1978" (GPO, 1980), table 1.

39. *Equal Opportunity: Second Annual Assessment of Programs* (Office of the Deputy Chief of Staff for Personnel, Department of the Army, 1978), pp. iii–iv. Appendix table B-13 shows the sources of commission for officers who entered the Army in fiscal 1981.

40. Herbert H. Denton and Barry Sussman, " 'Crossover Generation' of Blacks Expresses Most Distrust of Whites," *Washington Post,* March 25, 1981.

41. A comparison of expected earnings for college graduates is presented in chapter 4.

CHAPTER FOUR

BENEFITS VERSUS BURDENS

EQUITY ARGUMENTS in behalf of social representation within the armed forces are made usually from one of two distinct perspectives: (1) military service is a *burden* that should be borne equally by all members of society; and (2) the *benefits* associated with military service should be available to all individuals regardless of race, color, creed, national origin, or socioeconomic status.

Of course, changing times produce changing perspectives. During war, for instance, personal sacrifice and hardship define the burdens of defense. In times of peace, especially when peace is accompanied by high unemployment and a sagging economy, military service can mean a chance to be employed or to learn a skill and receive an education, particularly for the disadvantaged.

On the other hand, certain groups may see service in wartime as a benefit. For blacks and sansei during World War II, for example, combat duty meant the right to fight and acceptance as full citizens; exclusion from combat duty was a denial of citizenship and patriotism, and therefore of equality. During peacetime, military service is not invariably described in positive terms. Controversy over military representation has focused on disproportionate black enlistments, not because whites are being refused a fair share of the benefits, but because depressed minorities are viewed as *accepting* more than their fair share of the burdens in order to obtain the benefits.

Thus the way in which military service is perceived affects interpretations of recruiting results and prescriptions for parity. Ultimately, equity arguments hinge on the discernible distribution of rewards and reponsibilities, and this perception involves judgments that may differ across social and political lines. Some may see the traditional opportunities of military service as outweighing any negative aspects during times of peace. Consequently, the overrepresentation of disadvantaged minorities and the poor is viewed as social welfare and an equalization of social benefits, in much the same way the graduated income tax is seen as promoting equity.[1] The armed forces represent a chance to get ahead, an avenue for social and career mobility. The fact that the poor and depressed minorities enlist in disproportionate numbers is a healthy sign, an indication that these individuals can and will receive help. "It is a good thing and not a bad thing to offer better alternatives to the currently disadvantaged," Milton Friedman once observed.[2] Regardless of its shortcomings, many maintain, peacetime service makes the poor less poor and the unskilled skilled.

The notion of benefits and burdens is tied to the broader question of whether military service in a democratic society is a right or a duty. This is not a new issue; from the nation's beginnings national security has been a two-sided question. Should the American military be a professional force modeled after the armies of Europe, or should it be a nonprofessional force of citizen soldiers? Structure is intrinsically tied to recruitment. Massive citizen armies rely on the principle of universal obligation to service, usually through compulsion. Professional armies, on the other hand, are regular or career or, as some claim, all-volunteer armies. How to recruit men for military service has been debated in this country since the settlers fought with the Indians. That the controversy has "persisted through the whole history of the United States" is one gauge of its intractability.[3]

Recruitment and structure are likewise tied to the composition, or "descriptive" representation, of the armed forces. And while emphases

---

1. See, for example, Mark V. Pauly and Thomas D. Willett, "Who 'Should' Bear the Burden of National Defense?" in James C. Miller III, ed., *Why the Draft? The Case for A Volunteer Army* (Penguin Books, 1968), p. 63. Pauly and Willett also point out that "who shall serve?" and "who shall bear the burden of defense?" are two different questions; *all* bear the burden through payment of taxes but who serves is a matter of individual, voluntary choice (p. 68). An economist similarly observes that both rich and poor can reach higher welfare positions by this division and specialization of labor (though the disadvantaged have limited choices because of their economic status); see Steven L. Canby, *Military Manpower Procurement: A Policy Analysis* (D. C. Heath, 1972), p. 26.

2. Milton Friedman, "The Case for Abolishing the Draft—and Substituting for It an All-Volunteer Army," *New York Times Magazine,* May 14, 1967, p. 118. See also Robert D. Tollison, "Racial Balance and the Volunteer Army," in Miller, *Why the Draft?* pp. 148–58.

3. Russell F. Weigley, "Introduction," in John O'Sullivan and Alan M. Meckler, eds., *The Draft and Its Enemies: A Documentary History* (University of Illinois Press, 1974), p. xx.

have shifted over the years and certain democratic values have been redefined, the basic theme of debate about the proper composition of the American military has not changed.

The American Revolution was fought by a mixture of professional and citizen soldiers. A universal military obligation for nearly all males of appropriate age appeared in the statutes of all the British colonies (with the exception of Quaker Pennsylvania), and the obligation was enforced in the Indian wars. During the American Revolution, the newly independent states perpetuated the obligation.[4] At the end of the Revolution, though apparently favoring a small, professional army of competent regulars, George Washington nevertheless proclaimed a universal military obligation as the concomitant of the ballot, an idea that is the foundation of the modern mass army.[5] In his "Sentiments on a Peace Establishment," Washington wrote:

> It may be laid down as a primary position, and the basis of our system that every Citizen who enjoys the protection of a free Government, owes not only a proportion of his property, but even of his personal services to the defense of it, and consequently that the Citizens of America (with a few legal and official exceptions) from 18 to 50 Years of Age should be borne on the Militia Rolls, provided with uniform Arms, and so far accustomed to the use of them, that the Total strength of the Country might be called forth on a Short Notice on any very interesting Emergency.[6]

The rationale of a nation in arms advanced by Washington and his philosophical predecessors has supported conscription in democratic countries everywhere from revolutionary France to twentieth century United States. Yet Americans have usually resisted the idea of a standing army, seeing it as one of the vestiges of Old World monarchies and autocracies and a threat to basic liberties. In this country, the draft has received general popular acceptance for only relatively brief periods.[7] Compulsory service is seen as running "against the grain of the values

 4. Ibid., p. xv.
 5. Russell F. Weigley, *Towards an American Army: Military Thought from Washington to Marshall* (Columbia University Press, 1962), p. 12. Weigley notes that "professional" is used in this context to mean the opposite of "amateur"—that is, one competent in the art of fighting. "Professional" here thus does not refer to "profession" (the "military profession" or the "profession of officership") as distinguished from "trade." Ibid., p. 255, note 2.
 6. The text of Washington's response to Hamilton, "Sentiments on a Peace Establishment," appears in Walter Millis, ed., *American Military Thought* (Bobbs-Merrill, 1966), p. 23.
 7. In March 1863, two years after the Civil War began, Congress enacted the nation's first draft law. It was designed to stimulate the flow of Union troops in geographic areas that did not produce their quota of volunteers. The draft was invoked again in 1917 (World

of individualism and free choice that are far more deeply associated with the image of the United States among its citizens and in the world at large."[8] Nonetheless, as a military historian writes: "The thoughtful student of the history of the draft in the United States and of military history at large will find considerable substantiation for the contention that a citizens' army based on a universal obligation to serve is the most appropriate armed force for a democracy."[9]

Benefits of Military Service

In 1944 Gunnar Myrdal observed the importance of the armed forces as a source of employment for blacks:

In terms of economic value they offer some of the best opportunities open to many young Negro men. Food and clothing are excellent; the pay is higher than that in many occupations available to Negroes. And those conditions of employment are equal for Negroes and whites. A great number of poor Negroes must have raised their level of living considerably by entering the armed forces. It may be, also, that service in the Army and in the Navy, in many instances, will have a certain educational value that will make many Negroes better prepared for post-war employment.[10]

In recent years, upward of 80,000 young black men and women have entered the nation's armed forces annually and attained an economic status that many of them would find difficult to duplicate in the private sector. Military pay and benefits, job training and educational assistance, and social opportunities are particularly attractive to black youth; for many, in fact, the armed forces have provided their only opportunity for escape from ghetto life and from possible participation in the nation's underground economy.

The importance of the military as an employer of black youth is vividly illustrated in table 4-1, which shows that close to one out of every five black males born between 1957 and 1962 had entered the armed forces by September 1981, although only one of every nine whites had joined

---

War I). In 1940 (World War II) Selective Service legislation was once more enacted, and continued (with brief periods of interruption) until the last general draft call went out in December 1972. The so-called American military model that has emerged is that of a small, professional "caretaker" force in peacetime and a citizen army (of some form) during periods of war.

8. Weigley, "Introduction," p. xvii.

9. Ibid.

10. Gunnar Myrdal, *An American Dilemma: The Negro Problem and Modern Democracy* (Pantheon Books, 1972), vol. 1, p. 419.

Table 4-1. Military Participation Rates of Male Youths Born between 1957 and 1962, by Race and Education[a]

Percent

Educational level[b]	White[c]	Black	Total[d]
Non-high school graduate			
All youths	16.6	12.1	14.5
Qualified youths	39.0	135.7[e]	45.1
GED high school equivalency			
All youths	18.6	14.2	18.0
Qualified youths	25.5	37.6	27.0
High school graduate and above			
All youths	9.8	22.3	11.2
Qualified youths	10.2	33.7	12.2
Total			
All youths	11.5	18.2	12.3
Qualified youths	13.6	41.6	16.0

Sources: Statistics on qualified youths are derived from data in Department of Defense, "Profile of American Youth: 1980 Nationwide Administration of the Armed Services Vocational Aptitude Battery" (Office of the Assistant Secretary of Defense for Manpower, Reserve Affairs, and Logistics, March 1982); and special tabulations provided by the Office of the Secretary of Defense. Data on military personnel are from the Defense Manpower Data Center.

a. Participation rate is the percentage of male youths born between January 1, 1957, and December 31, 1962, who enlisted in the military (for the first time) between July 1973 and September 1981. Participation rates are shown for two base populations: (1) all male youths within the racial-ethnic and educational category; and (2) all male youths who would be expected to *qualify* for enlistment under 1981 aptitude test standards (by racial-ethnic and educational category). The cross-sectional participation rates *understate* the true percentage of male youths who join the military since they do not include those who (a) enlisted after September 30, 1981, and (b) entered officer programs. Estimates of the number of youths qualified for military service were calculated from the results of the Profile of American Youth (administration of the Armed Services Vocational Aptitude Battery to a national probability sample in 1980) and the 1981 education-aptitude standards used by the armed services. (Eligibility for enlistment would also depend on other factors—including medical and moral requirements.)

b. For military personnel, education at time of entry (and initial qualification) into service. Approximately 1 percent of the male youth population could not be identified on the basis of education; and 1 percent of military personnel could not be identified on the basis of racial-ethnic group. These unknown cases were not included in the calculations of participation rates.

c. White category includes all racial-ethnic groups other than black or Hispanic.

d. Total includes all racial and ethnic groups.

e. This figure reflects the fact that during the fiscal 1976–80 period the armed services unknowingly accepted volunteers who did not meet eligibility standards because of errors in test calibration. Since these errors affected principally non-high school graduates with low aptitude scores, the services enlisted many more black male dropouts than would have been qualified in the relevant population group.

by then. The contrast is even sharper when account is taken of the fact that blacks are less likely to qualify for enlistment. For example, by conservative estimate, almost *42 percent* of all potentially qualified black males had enlisted by the end of fiscal 1981. The comparable "participation rate" for potentially qualified white males was less than 14 percent.

Perhaps an even more revealing aspect of youth participation is that potentially qualified young men who do not have a high school diploma or equivalency certificate—regardless of race—find military service an especially appealing job or education alternative. As shown in table 4-1, almost half of all high school dropouts who could probably pass the military's aptitude test standards had enlisted, and more than one out of four qualified, General Educational Development (GED) recipients had made the same choice. In fact, the image of the armed forces as a second chance, a place of equal acceptance and involvement despite prior social disadvantage or preexisting handicap, has helped to make the military a traditional channel for social mobility. The participation rates displayed in the table tend to confirm that both the image and the promise of opportunity are still quite strong.

Surveys conducted throughout the history of the all-volunteer military likewise attest to the lure of being "all that you can be" (as the Army recruiting jingle promises) for blacks and whites alike.

In the 1979 Survey of Personnel Entering Military Service, enlistees were asked to specify their main reasons for volunteering. The most popular—selected by over nine out of ten new recruits regardless of race—was "better myself in life." Personal betterment was chosen more often than any other alternative by blacks (40.1 percent) as *the most important* reason for joining the military, although white recruits most often (34.9 percent) selected "get training for a civilian job." (Almost 34 percent of all white recruits, however, selected "better myself in life" and 26 percent of blacks chose "get training for a civilian job" as the most important reason for enlisting.) About one out of ten blacks and one out of fourteen whites indicated that to "get money for college education" was their principal motive—whereas one out of ten whites and one out of fourteen blacks claimed that they enlisted primarily to "serve my country."[11]

Though black youth may find different features of the military the most appealing, the common factor that influences its overall attractiveness, particularly to young black males, is the dismal civilian labor market that confronts them. They are more likely to be unemployed and, when employed, to earn less on the average than their white counterparts.

Black Youth Unemployment

Despite the attention that has been directed toward the issue of youth unemployment, there is still no consensus on the nature and dimension of the problem. There does appear to be, however, wide agreement on

11. Results of the 1979 Department of Defense Survey of Personnel Entering Military Service; data obtained from the Department of Defense, Defense Manpower Data Center. Complete survey results are shown in appendix table B-14.

Table 4-2. Unemployment Rates, by Race and Age, Selected Years, 1955–81
Percent

Race and age	1955	1965	1973	1978	1981
All workers 25 and over	3.6	3.2	3.1	4.1	5.4
White males					
16–19	11.3	12.9	12.3	13.5	17.9
16–17	12.2	14.7	15.2	16.9	19.9
18–19	10.4	11.4	10.0	10.8	16.4
20–24	7.0	5.9	6.6	7.7	11.6
Black males[a]					
16–19	13.4	23.3	30.7	36.7	40.7
16–17	14.8	27.1	35.7	43.0	43.2
18–19	12.9	20.2	23.0	32.9	39.2
20–24	12.4	9.3	13.2	21.0	26.4

Sources: 1955 and 1956, Bureau of Labor Statistics, *Handbook of Labor Statistics;* other years, unpublished figures from BLS, Current Population Survey.
a. Figures for 1955 and 1965 include all nonwhite males.

several points: (1) unemployment rates for youths are higher than those for adults; (2) there has been a secular growth in teenage unemployment; and (3) the growth in black youth unemployment has outrun the growth in white youth unemployment.[12]

The unemployment rate for the American labor force as a whole in December 1981 was 8.9 percent. At the same time, the unemployment rate for black teenagers, according to the Bureau of Labor Statistics, was 39.6 percent, but for many black youths in large urban areas, unofficial estimates placed the effective rate closer to 60 percent.[13] In contrast, the rate for white teenagers was 19.3 percent. This pattern is roughly consistent in each age group, with black youth unemployment rates more than twice and as much as three times as large as comparable white rates. As table 4-2 shows, this disparity is a continuation of the postwar trend of widening differences between the unemployment rates of black and white youth: while the unemployment rates of white males in all age groups in 1981 were higher than they were in 1955, the sharp increase in the jobless rates of young black males over the period is conspicuous.

Overall, the teenage unemployment rate in recent years has averaged

12. For an excellent discussion, see Michael L. Wachter, "The Dimensions and Complexities of the Youth Unemployment Problem," in Bernard E. Anderson and Isabel V. Sawhill, eds., *Youth Employment and Public Policy* (Prentice-Hall, 1980), p. 33.

13. The unemployment rate is defined as the number of unemployed divided by the number in the labor force. The unemployed are all those who are out of a job but are "actively seeking work."

Table 4-3. **Employment–Population Ratios, by Race and Age, Selected Years, 1955–81**

Race and age	1955	1965	1973	1978	1981
White males					
16–19	52.0	47.1	54.3	56.3	51.3
16–17	42.2	38.0	44.7	46.0	41.2
18–19	64.2	58.3	65.1	67.2	61.4
20–24	80.4	80.2	80.2	80.6	77.0
Black males[a]					
16–19	52.7	39.4	32.8	28.5	24.6
16–17	41.1	28.8	20.8	18.3	16.5
18–19	66.0	53.4	47.0	40.2	33.3
20–24	78.6	81.6	72.6	62.2	58.3

Sources: Same as table 4-2.
a. Figures for 1955 and 1965 include all nonwhite males.

five times the rate for the labor force aged twenty-five or older, and the rate for young adults has averaged two and a half times that for workers over twenty-five. But the unemployment rates for young black adults have been roughly five times and for black teenagers approximately seven times the rate for all males in the over-twenty-five age category.

While the unemployment statistics for black youth are alarming enough, they fail to account for the large numbers of young blacks who would prefer to work but instead withdraw from the labor force out of frustration and join the ranks of the "discouraged workers"[14] or for those who join the underground economy and are not counted in the national unemployment statistics. To the extent that sizable numbers of young blacks are represented in these two groups, unemployment rates and other labor market measures for black youth understate the severity of the employment difficulties that beset them.

A different, though complementary, perspective on the youth unemployment situation is provided by an examination of employment–population ratios, which are simply measures of the number of workers as a percentage of the population. In addition to the persistent increase (in both absolute and relative terms) in unemployment rates for young black males, further indications of a temporal deterioration in their employment situation are evident in the substantial decline in their employment–population ratios between 1955 and 1981 (see table 4-3).

14. For more detail on the discouraged worker effect, see Francine D. Blau, "Youth Participation Rates and the Availability of Jobs," in *Supplementary Papers from the Conference on Youth Unemployment: Its Measurement and Meaning* (Department of Labor, 1978), pp. 56–77; and Alfred Tella, "Labor Force Sensitivity to Employment by Age, Sex," *Industrial Relations*, vol. 4 (February 1965), pp. 69–83.

Whereas the ratios for both black and white teenagers were roughly the same in 1955, by 1981 the ratio for white male teenagers had declined slightly while the ratio for black teenagers had fallen precipitously. Young adult blacks also experienced a substantial decline in their employment–population ratio between 1955 and 1981 as the ratio for their white counterparts fell slightly. Thus employment for young white males grew faster over the period than their population, which was also growing rapidly. For black youths, on the other hand, employment grew at a slower rate than their population. Taken together, the declining employment–population ratios and the deterioration in relative unemployment rates "indicate that the youth labor market problem is concentrated among blacks."[15]

Even more controversial than the dimensions of the problem are its causes.[16] One of the most commonly mentioned but unverifiable factors is labor market discrimination. Employers often cite the lack of skills and experience of the young as a major reason for not hiring them. Moreover, the majority of the young land entry-level or part-time jobs that characteristically offer little upward mobility or provide no opportunity to develop valuable job skills. Thus many find themselves subjected to a kind of "Catch-22"—they cannot be hired for a job offering career advancement without prior work experience, but must work at such a job in order to gain the needed experience. Fortunately for some, this circumstance is mitigated by the various youth programs that concentrate specifically on the acquisition of job skills.

Race discrimination figures prominently in explaining the high unemployment rate of young blacks. Economists have acknowledged the role race may play in the labor market, and theories of discrimination range from an explanation that employers have a "taste" for discrimination[17] to a more recent view—referred to as "statistical theory

15. Wachter, "Dimensions and Complexities," p. 58. Wachter is quick to point out, however, that the extent of the problem is difficult to assess. "The declining employment to population ratios and increasing unemployment rates have taken place in a period of rising black youth school enrollment rates and increasing relative wages for black youth. These developments may suggest that, within the black youth group, there is an increasing dispersion of labor market opportunities and performance" (p. 61).

16. An interesting framework for discussion is provided in Richard B. Freeman, "Why Is There a Youth Labor Market Problem?" in Anderson and Sawhill, eds., *Youth Employment and Public Policy,* pp. 11–12.

17. According to this argument, "If an individual has a 'taste for discrimination' he must act *as if* he were willing to pay something, either directly or in the form of a reduced income, to be associated with some persons instead of others. When actual discrimination occurs, he must, in fact, either pay or forfeit income for this purpose." Gary S. Becker, *The Economics of Discrimination* (University of Chicago Press, 1957), p. 6.

of discrimination"[18]—that discriminatory behavior is based on the average performance of the racial group to which an individual belongs and that race is used by the employer as a cheap source of information about the applicant (in this case, minority youth). Since race interacts with other factors, such as inadequate (or poor) education and low socioeconomic status, the isolation and measurement of racial discrimination is a difficult task.

Regardless of which comparative labor market indicators are used or what factors may be contributing to the situation, the youth unemployment problem is of unquestionably large proportions overall and is even more pronounced for black males. It explains why so many have turned to the armed forces.

Relative Earnings

Young blacks who are able to find civilian employment can expect to earn less on the average than whites of similar age and education. The differences are shown in figure 4-1, which compares the mean earnings of white and black males who have completed high school but not college. When contrasted with the average earnings for military enlisted personnel, it becomes evident why military pay scales, *which are the same regardless of race,* are relatively more attractive to blacks than to whites.[19] While efforts to increase the number of blacks in the officer corps do not show it, a sharper contrast in average annual earnings (shown below in dollars) for 1979 was found among college graduates:[20]

	Age group		
	25–34	*35–44*	*45–54*
White civilians	19,452	28,532	30,355
Black civilians	17,928	18,848	21,720
Military officers	21,941	30,297	34,160

18. See, for example, Edmund S. Phelps, "The Statistical Theory of Racism and Sexism," *American Economic Review,* vol. 62 (September 1972), p. 659; and Marvin M. Smith, "Towards a General Equilibrium Theory of Racial Wage Discrimination," *Southern Economic Journal,* vol. 45 (October 1978), pp. 458–68.

19. All members of the armed forces who hold the same rank and who have served the same amount of time receive the same basic pay, regardless of sex, race, or other dissimilarities. To the extent, however, that race enters into occupational differences and hence into bonus eligibility and promotion opportunity, the average earnings profiles will differ.

20. Based on unpublished data provided by the Bureau of the Census. Earnings data for military officers are authors' calculations based on data provided by the Defense Manpower Data Center. Broader age groups are used for college graduates than for high school graduates since the population bases for black college graduates in more limited age groupings are too small to be reliable.

Figure 4-1. Comparison of Mean Annual Earnings of Male Civilians, by Race and Age Group, and of Military Enlisted Personnel, by Age Group, 1979[a]

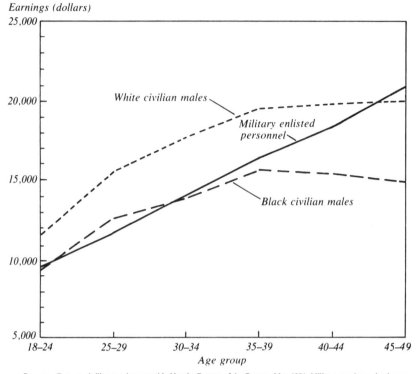

Sources: Data on civilian earnings provided by the Bureau of the Census, May 1981. Military earnings—basic pay, subsistence and quarters allowances, and tax advantage—provided by the Department of Defense, May 1981.

a. Includes civilians with at least four years of high school but less than four years of college employed full time.

The "Bridging" Factor

For minority youths who choose not to remain in the armed forces on a career basis, military service still affords many the chance to prepare themselves for more rewarding civilian opportunities. The mechanisms by which military service can alter the status of young blacks who have been isolated from the mainstream of American life are part of a "bridging environment," in which the individual acquires new skills and abilities to help him in his civilian career.[21] The unfavorable conditions for

21. Harry L. Browning, Sally C. Lopreato, and Dudley L. Poston, Jr., "Income and Veteran Status: Variations Among Mexican Americans, Blacks and Anglos," *American Sociological Review*, vol. 38 (February 1973), p. 76.

mobility that affect many young blacks who grow up in inner-city ghettos are altered by the military experience in several ways.

First, military service provides many youths with the initial opportunity to loosen their tight bonds to family and neighborhood groups and to weaken ethnic boundaries, thus fostering greater independence.[22]

Second, as the nation's largest single vocational training institution, the armed services offer an opportunity to acquire skills and knowledge that not only enable service members to carry out their military duties but also, in many instances, prepare them for more productive careers when they leave the service. In fiscal 1982, for example, about 386,000 recruits expected to enter about 750 different initial skill-training courses, and some 44 percent of them also were offered advanced training.[23] These courses encompass a wide range of occupations; some, such as combat training, have little transferability to civilian life, while others, such as law enforcement and mechanical and electronics training, make military service a path to a more productive civilian career. Also, their active service provides military personnel with postservice educational assistance, and a variety of educational programs are available to them while they are still on active duty.

Third, military service gives ethnic minorities the opportunity to become familiar and learn to cope with large bureaucratic structures similar to those encountered in the nonmilitary sector, an important attribute, it is argued, for converting education into dollars of earnings.[24]

Some contend, however, that the benefits of military service as a bridge between the ghetto and a successful civilian career have been overstated. The *New York Times* put it bluntly:

The myth that Army service can help by training deprived blacks for civilian jobs is little borne out by the facts. Usable civilian skills are least likely to be acquired in the ground combat forces, where most blacks are, while they are under-represented in the technical and support services, a vastly better training ground.[25]

The jobs for which black veterans are found to earn more than black nonveterans, so this argument goes, are largely restricted to the "occu-

---

22. Ibid.
23. Department of Defense, "Military Manpower Training Report for FY 1982" (March 1981), pp. V-4, V-7.
24. Sally C. Lopreato and Dudley L. Poston, Jr., "Differences in Earnings and Earnings Ability Between Black Veterans and Nonveterans in the United States," *Social Science Quarterly,* March 1977, p. 759.
25. *New York Times,* February 5, 1975.

Table 4-4. Distribution of Male Enlisted Personnel, All Services, by Major Occupational Category and Race, September 1981

Major occupational category[a]	White		Black	
	Number	Percent	Number	Percent
White collar	429,516	43.3	138,346	43.4
Technical workers[b]	291,409	29.4	68,183	21.4
Clerical workers[c]	138,107	13.9	70,163	22.0
Blue collar	562,797	56.7	180,327	56.6
Craftsmen[d]	313,806	31.6	70,176	22.0
Service and supply workers	92,138	9.3	42,152	13.2
General military skills, including combat	156,853	15.8	67,999	21.3
Total[e]	992,313	100.0	318,673	100.0

Source: Derived from data provided by Defense Manpower Data Center.
a. Categories are based on the Defense Department occupational classification system.
b. Includes "electronic equipment repairmen," "communications and intelligence specialists," "medical and dental specialists," and "other technical and allied specialists" categories.
c. Includes the "functional support and administration" category.
d. Includes "electrical/mechanical equipment repairmen" and "craftsmen" categories.
e. Totals do not include enlisted personnel in a nonoccupational status (for example, in training or in hospital).

pational categories least affected by career continuity (clerical workers; sales workers; operators; service workers; and laborers)."[26]

The occupational data presented in chapter 3 suggest that blacks are disproportionately represented in the softer skills. This is confirmed in table 4-4, which aggregates military occupations into a civilian-oriented occupational classification. Although black males in the enlisted ranks are just as likely as whites to hold a white-collar job, they are more concentrated in clerical positions and whites are more apt to be in technical billets. Also, among male enlisted personnel with blue-collar skills, whites are more likely to be trained as craftsmen and blacks as

---

26. Browning and others, "Income and Veteran Status," p. 77. The research community is not in unanimous agreement that military training is useful in the civilian labor market. Some have pointed out that, although "more than three-fourths of the job specialties available to enlistees have direct civilian counterparts," the transferability of military training to the civilian sector seems to depend in large measure on the nature of the occupation. Moreover, much of the empirical research shares the same shortcoming: a failure to distinguish earnings effects attributable to military service from those resulting from innate differences. See, for example, Eva Norrblom, *An Assessment of the Available Evidence on the Returns to Military Training*, R-1960-ARPA (Santa Monica: Rand Corp., 1977), pp. iv, v; Eugene L. Jurkowitz, "An Estimation of the Military Contribution to Human Capital: Appendix E," in Paul A. Weinstein, ed., *Labor Market Activity of Veterans: Some Aspects of Military Spillover*, OEG2-6-062198-1955 (University of Maryland, Department of Economics, 1969), pp. E1–E84; and William Mason, "On the Socioeconomic Effects of Military Service" (Ph.D. dissertation, University of Chicago, 1970).

service and supply workers or in general military skills, including combat. All in all, while the military clearly provides employment opportunities for black youths, whites tend to acquire training and skills that put them in a position to compete for better jobs in the civilian sector.

Burdens of Military Service

Many Americans view military service not as a benefit, but as a burden, oppressive in peacetime because its coercive discipline runs counter to the individualistic tradition of American society, but particularly onerous in wartime as the risks of death or injury multiply. The prospect that these risks may be borne disproportionately by minority members of the nation's armed forces causes concern, not only as a matter of egalitarian principle, but also because of the potential for promoting social divisiveness at a juncture when the nation can least afford it. Moreover, just the prospect might somehow influence the range of military options available to national authorities.

Who among the nation's citizenry should serve, and perhaps die, in the nation's defense has been a controversial issue since the beginning of democratic societies, tied into the debate over whether service in the armed forces is a right or a duty. For most of its history, the United States has relied on both concepts. Before World War II the peacetime military consisted of small, cadre-type volunteer forces that were expected to serve as the nucleus around which mobilization of civilian soldiers would take place.

After the war the nation, to support its cold war strategy, adopted peacetime conscription to field the largest standing army in its history. Though the military draft could hardly be considered a model of equity, dissent was minor and sporadic.[27] In 1951 the fairness of student deferments was debated and a policy adopted that for the next two decades was to protect college students from the draft. Interracial equity in reserve manpower policy became an issue in 1955 but was not resolved. Men over the age of twenty-five were exempted from induction in 1956 on the ground that older, more settled men should not be taken from their careers and families. In 1958 some worried that the highly selective entry standards discriminated against youth from lower socioeconomic

---

27. James M. Gerhardt, *The Draft and Public Policy: Issues in Military Manpower Procurement, 1945–1970* (Ohio State University Press, 1971), pp. 359–60.

Table 4-5. Army Combat Deaths in Vietnam, by Race, 1961–72

Year or period	Number of deaths		Blacks as percent of total
	Blacks	Total	
1961–64	12	185	6.5
1965	186	898	20.7
1966	639	3,073	20.8
1967	730	5,443	13.4
1968	1,220	9,333	13.1
1969	772	6,710	11.5
1970	318	3,508	9.1
1971	110	1,269	8.7
1972	13	172	7.6
1961–66	837	4,156	20.1
1967–72	3,163	26,435	12.0
1961–72	4,000	30,591	13.1

Source: Department of the Army, "Information Paper: Blacks in Vietnam Conflict," March 3, 1977.

strata, but the standards remained intact. And the partial mobilization of reserves during the Berlin buildup in 1961 prompted discussion about inequities of selection and discharge procedures. But it was not until the mid-1960s and the onset of the longest and most difficult war in the nation's history that political apathy toward equity-related issues began to disappear and the protective and exclusionary features of the Selective Service System were seriously questioned.

By 1967 the notion of "numerical fairness" had become a heated issue as it was suggested that blacks were shouldering a disproportionate burden of the war. Between 1961 and 1966, when blacks constituted approximately 11 percent of the general population aged nineteen to twenty-one, they accounted for one out of every five Army combat deaths (see table 4-5). Although this mortality rate was more or less in proportion to the number of blacks in combat units, civil rights spokesmen had the evidence to claim that the military was unjustly using black youth as "cannon fodder for a war directed by whites." According to Stokely Carmichael, it was "clear that the [white] man is moving to get rid of black people in the ghettoes."[28] Martin Luther King, Jr., advocated a boycott of the war by blacks, claiming that they were "dying in disproportionate numbers in Vietnam"; the heads of the Congress of

28. Sol Stern, "When the Black G.I. Comes Home From Vietnam," in Jay David and Elaine Crane, eds., *The Black Soldier: From the American Revolution to Vietnam* (Morrow, 1971), p. 221.

Racial Equality (CORE), the National Urban League, and the Student Non-Violent Coordinating Committee (SNCC), and other civil rights leaders spoke of the "imbalance of black Americans in the war in Vietnam," the "racist policies of the Selective Service System," and the "disproportionate percentage of the burdens" placed on young black men.[29]

In February 1967, with the release of a report by the National Advisory Commission on Selective Service (the Marshall Commission), the charges of discrimination by civil rights leaders were validated. The commission based its work on the premise that specified groups (racial, social, economic) should bear the risk (or incidence) of death in war and the responsibilities of service in peacetime only in *rough proportion to that group's percentage in society*. Citing evidence of the "Negro's over-representation in combat," the commission concluded that "social and economic injustices in the society itself are at the root of the inequities which exist."[30]

Press accounts indicate that in reaction to the disproportionate death rates in 1965 and 1966, "the Pentagon ordered a cutback in front-line participation by Negroes."[31] An Army general was quoted as saying, "We deliberately spread out Negroes in component units at a ratio pretty much according to the division total. We don't want to risk having a platoon or company that has more Negroes than whites overrun or wiped out."[32] The change was quite dramatic; as table 4-5 indicates, the proportion of combat fatalities among black troops dropped in the following year and continued to decrease for the remainder of the war. As it turned out, blacks suffered 13 percent of the combat deaths during the entire wartime period (1961–72), slightly above the percentage of blacks in the relevant-aged population, and coincidentally almost exactly the same percentage as that of blacks in the Army's enlisted ranks.

The concern, however, was confined to a small but vocal group of black leaders; there was little evidence that black troops themselves were worried about it. In 1966—when the proportion of black casualties

29. See Tollison, "Racial Balance and the Volunteer Army," p. 148; Paul T. Murray, "Local Draft Board Composition and Institutional Racism," *Social Problems*, vol. 19 (Summer 1971), pp. 129–37; and Ulysses Lee, "The Draft and the Negro," *Current History*, vol. 55 (July 1968), p. 47.

30. *In Pursuit of Equity: Who Serves When Not All Serve?* Report of the National Advisory Commission on Selective Service (Government Printing Office, 1967), p. 10.

31. Thomas A. Johnson, "The U.S. Negro in Vietnam," *New York Times*, April 29, 1968.

32. *U.S. News & World Report*, August 15, 1966, p. 62.

in Vietnam peaked—black GIs who had completed their first term of service were reenlisting in the Army at a rate over three times that of their white peers (66.5 and 20.0 percent).[33]

Regardless of the size of the dissenting group, it is the lingering memory of the racial divisiveness of that period that today prompts some observers to fear the possible social consequences of U.S. involvement in another military conflict. If that happened, the 20 percent casualty rate of blacks that provoked charges of racial genocide in the mid-1960s could appear small. Even if black casualties were just in proportion to blacks in the Army's enlisted ranks and, coincidentally or not, to the proportion with combat specialties, one out of every three Army combat deaths in the early stages of a conflict would be that of a black soldier. Unattractive though that prospect may be, it probably represents the lower bound; black casualties in the opening days of any military engagement are likely to be heavier than 33 percent in the Army, depending on the specific units involved in combat.

Should the Army's 2nd Infantry Division become involved in hostilities on the Korean peninsula, black soldiers would be likely to suffer 40 percent of the early casualties, based on the racial mix in the division (see table 4-6). The proportion could be larger, depending on which units of the 2nd Division were engaged. As of December 1980, for example, the proportion of enlisted blacks in its five infantry battalions ranged from about 38 to 46 percent; the proportion was 38 percent in both its armored battalions, and from 44 to 51 percent in its four artillery battalions.[34]

An outbreak of hostilities in Central Europe, on the other hand, would involve the 3rd Armored Division, whose enlisted ranks are about 34 percent black and whose combat battalions have a similar mixture (see appendix table B-15). Relative casualty rates that would accompany the commitment of the U.S. Rapid Deployment Force would depend on which units were used. If the 82nd Airborne Division was involved—as it surely would be—black casualties would be relatively low. Even though the proportion of blacks in the enlisted ranks of the 82nd is roughly 26 percent, none of its infantry battalions exceed 23 percent and one is as low as 16.4 percent black. The 101st Airborne Division, which is also earmarked for the rapid deployment mission, is about 34 percent black, with one of its infantry battalions close to 38 percent black.

33. *New York Times,* April 29, 1968.
34. See appendix table B-15 for detailed breakdowns of minority content by battalions.

Table 4-6. Blacks as a Percentage of Total Enlisted Personnel Assigned to Army Divisions and Selected Brigades and Regiments, September 1980

Location and organization	Percent black
United States	
82nd Airborne Division	26.0
101st Airborne Division (Air Assault)	33.8
2nd Armored Division	35.8
1st Cavalry Division	38.2
1st Infantry Division (Mechanized)	32.2
4th Infantry Division (Mechanized)	25.9
5th Infantry Division (Mechanized)	40.8
7th Infantry Division	39.8
9th Infantry Division	28.2
24th Infantry Division	40.5
25th Infantry Division	29.6
6th Air Cavalry Combat Brigade	34.1
194th Armored Brigade	29.3
172nd Infantry Brigade	24.3
197th Infantry Brigade	51.8
3rd Armored Cavalry Regiment	26.8
Europe	
1st Armored Division	33.0
3rd Armored Division	34.2
3rd Infantry Division (Mechanized)	34.0
8th Infantry Division (Mechanized)	34.3
Berlin Brigade	31.6
2nd Armored Cavalry Regiment	29.9
11th Armored Cavalry Regiment	30.1
Korea	
2nd Infantry Division	41.1
Panama	
193rd Infantry Brigade	35.1

Source: Department of the Army, May 1981.

The heaviest black casualties would be likely to occur upon the commitment to combat of the 197th Infantry Brigade, whose enlisted component is over 50 percent black. As appendix table B-15 shows, its infantry battalions are 51 and 55 percent black, its artillery battalion is 60 percent black, and one of its artillery batteries runs as high as 66 percent black.

The Marine Corps, too, faces the prospect of disproportionate casualties among its minority members. Whereas roughly 22 percent of all enlisted marines are black, they constitute about 29 percent of division strength distributed as follows: the First Marine Division, headquartered at Camp Pendleton, California, is 25 percent black, the Second Division, at Camp Lejeune, North Carolina, is 33 percent black, and the Third Division, whose home base is on Okinawa, is about 30 percent black. Within each division, there is little variation among the infantry and artillery battalions, with the smallest proportion of blacks found in tank and assault amphibious battalions (see appendix table B-16).

Should a conflict be protracted, casualty rates would tend to become more representative as reserve units were mobilized and conscription put into effect. How much this would shift the burden would depend on the particular scenario and on assumptions about mobilization and conscription. One scholar contends that a volunteer military that enlists largely from the lower socioeconomic strata actually insulates the poor and disadvantaged from the horrors of war and subjects the more privileged to a greater wartime risk.[35] The very fact that the less privileged choose to enter the military in peacetime, so this argument goes, means that in a war requiring mobilization the more privileged social classes would be drafted disproportionately into the segments of the armed forces that need to be expanded—the ground combat forces. In wartime, it is the combat arms that require most additional manpower, and better-educated draftees and volunteers would find fewer noncombat spaces open to them. On the other hand, careerists, even those with combat skills, would probably be remote from direct fighting, in higher-ranking cadre and staff positions. According to this view, "a strong case can be made that peacetime voluntarism and an equal-probability wartime draft [such as a lottery] in the modern dual nuclear-conventional military would disproportionately endanger the *more* privileged, and not excessively burden the poor."[36] The strength of the case, however, rests on the fragile assumptions that an equal-probability draft with few exemptions and deferments could be implemented and that a cross section of American youth would wind up being trained for combat. In light of past history, a healthy measure of skepticism is warranted on both counts.

In any case, since the end of the Vietnam War, misgivings about disproportionate black casualties have been voiced more by whites than by blacks. The potential problem was raised in 1970 by the President's Commission on an All-Volunteer Armed Force:

Members of both the white and Negro communities have expressed concern that the all-volunteer force might fill its enlisted ranks with the poor and the

---

35. Canby, *Military Manpower Procurement*, p. 26.
36. Ibid.

black. . . . This will result, according to some, in the black and the poor bearing a disproportionate share of the burden of defense.[37]

The commission, however, dismissed the prospect on the basis of their estimate that blacks would not be likely to make up more than 15 percent of the total male enlisted force and 19 percent of the Army enlisted ranks. Besides, the commission concluded, "blacks who join a voluntary force presumably have decided for themselves that military service is preferable to the other alternatives available to them. . . . Denial of this opportunity would reflect either bias or a paternalistic belief that blacks are not capable of making the 'right' decisions concerning their lives."[38]

Among the members of the unusual coalition of northern liberals and southern conservatives who opposed the end of conscription, Senator Edward Kennedy was one of the most vocal. In the debate over the end of conscription, he stated that "it is inequitable to permit the risks of battle to fall only on those less affluent Americans who are induced to join the army by a payraise."[39] Senator Claiborne Pell echoed the sentiment: "Too large a proportion of this burden would, I believe, be carried by our black citizens."[40]

But some blacks are cynical about the seemingly altruistic motives of whites who express such concern. Among them is Representative Shirley Chisholm:

All this talk about a volunteer Army being poor and black is not an indication of "concern" for the black and the poor, but rather of the deep fear of the possibility of a Black army. Individuals who are upset over Black Power rhetoric really shudder at the idea of a whole army of black men trained as professional soldiers.[41]

37. *The Report of the President's Commission on an All-Volunteer Armed Force* (Macmillan, 1970), p. 138.
38. Ibid., pp. 15, 16.
39. *Congressional Record,* vol. 116, August 25, 1970, p. 29972.
40. Ibid., p. 29973.
41. Testimony of Shirley Chisholm before the House Armed Services Committee, March 11, 1971, in *Congressional Digest,* vol. 50 (May 1971), p. 156. Chisholm would have had cause to single out Congressman Paul N. (Pete) McCloskey, Jr., for his comments on the issue in 1978. In speaking to a group of Stanford University alumni, McCloskey reportedly said: "What we're seeing now is a growth in minorities in U.S. forces. It's dangerous to have an all-black Army. For example, the Mafia was able to maintain its political and financial success by infiltrating the New York Police Department" (quoted in the *Stanford Daily,* October 23, 1978). Needless to say, the remarks stirred a good deal of controversy and in a subsequent "clarification" of his statements, the congressman argued for an Army that is reasonably representative of the people. "No one group—whether they be all-whites, all-blacks or all-Irish—should be carrying all the weapons in the military" (*Stanford Daily,* November 3, 1978). "McCloskey addresses the unspoken fears

By 1974 it was obvious that growth in the percentage of minorities in the Army, which was discernible even before the end of conscription, would be likely to continue at a rate in excess of earlier predictions. Two scholars were among the first to broach the question: "What can be done to deal with the possibility of a concentration of casualties in one segment of society?"[42] Their question was largely ignored because, first, with the Vietnam experience still fresh, the possibility that the nation would soon get involved in a new military action seemed remote, and second, the few black leaders who considered the issue seemed to place greater emphasis on the positive aspects of the military employment opportunities. Congressman Ronald V. Dellums said:

> Black volunteers understand what joining the military means. If, through exercise of free choice by individuals, there are [proportionately] more blacks in the service than in the population, we should expect a proportionately greater sacrifice. The whole idea of a volunteer Army is that the individual will take this risk and this responsibility on by his or her free choice.[43]

And Vernon Jordan pointed out the following asymmetry:

> It is interesting that while American servicemen were fighting and dying in Vietnam there was little concern about the disproportionate numbers of blacks in front-line combat units and the casualty lists. Criticisms of the armed forces should be based on real issues, not on false racial concerns.[44]

But black leaders have not spoken with one voice on the issue. Congressman (D.C. Delegate) Walter E. Fauntroy, for example, took a different view in justifying a 1979 meeting between Ambassador Andrew Young and representatives of the Palestine Liberation Organization: "Based on history, any U.S. military involvement in a Mideast war would again mean a disproportionate impact on black military personnel and their families."[45]

---

of a few," according to one editorial, "when he explicitly likens recruitment of blacks into a volunteer military to organized crime infiltration of the New York City police department" (*Inquiry*, April 2, 1979).

42. Morris Janowitz and Charles C. Moskos, Jr., "Racial Composition in the All-Volunteer Force," *Armed Forces and Society*, vol. 1 (November 1974), p. 110.

43. Ronald V. Dellums, "Don't Slam Door to Military," *Focus*, vol. 3 (June 1975), p. 6.

44. *Washington Star*, July 13, 1975. Some white conservatives, meanwhile, echoed a similar theme. A Hoover Institution scholar wrote that "the so-called fears [of an all-black armed force]—though unjustified in the first place—of some of our more liberal social commentators have proved to be unfounded.... The reasons for having apprehensions about 'too many blacks' in the armed forces have always been a little vague, or perhaps we should say they have been stated vaguely." See Martin Anderson, "The All-Volunteer Force Is Working," *Commonsense*, vol. 2 (Fall 1979), p. 9.

45. Letter to the Editor, *New York Times*, August 31, 1979.

How black Americans would actually react to heavy military casualties, particularly those of their most promising young men, is hard to predict. "It is naive, if not duplicitous," according to one scholar, "to state that disproportionately high black casualties will have no or only minor consequences on the domestic political scene."[46] "In the emotional climate aroused by combat deaths," observes another, "attitudes and feelings are subject to volatile shifts and may suddenly crystallize into adamant opposition to the military action, especially under the stimulation of concentrated and graphic television coverage of a highly telegenic 'issue.' "[47]

THE BENEFITS versus burdens issue confronts us with a real policy dilemma, the contours of which have been succinctly drawn by Charles Moskos:

It is a commentary on our nation that many black youths, by seeking to enter and remain in the armed forces, are saying that it is even worth the risk of being killed in order to have a chance to learn a trade, to make it in a small way, to get away from a dead-end existence, and to become part of the only institution in this society that seems really to be integrated.[48]

The dilemma is shaped by perceptions of equity, which are greatly influenced by the assumed ratio of benefits to burdens in the military service. When the burdens of enlistment are seen to outweigh the benefits, attention is focused on social class distinctions, and any overrepresentation of individuals from the lower social strata is perceived as evidence of systemic inequity. Where the benefits of military service outweigh the perceived burdens, however, it has been suggested that true social equity can be achieved by the overrepresentation of the disadvantaged poor and racial minorities.

46. Charles C. Moskos, Jr., "Symposium: Race in the United States Military," *Armed Forces and Society*, vol. 6 (Summer 1980), p. 589.
47. Robert K. Fullinwider, "The AVF and Racial Balance," Working Paper MS-1 (University of Maryland, Center for Philosophy and Public Policy, March 1981), p. 16.
48. Charles C. Moskos, Jr., *The American Enlisted Man: The Rank and File in Today's Military* (Russell Sage Foundation, 1970), p. 133.

CHAPTER FIVE

RACIAL COMPOSITION AND NATIONAL SECURITY

THE ISSUE discussed in the previous chapter, though important in its own right, barely touches on the preeminent question of military capability. The chief function of the military establishment is to defend the republic, and whether the racial makeup of the armed forces has any bearing on their capacity to fulfill that mission must be examined.

This is a sensitive question, steeped in suspicion and invariably couched in innuendo; that discussion of it has been largely emotional is not surprising. To some, merely raising the question implies the existence of an inverse relationship between minority content and military effectiveness.[1] And the conceptual difficulties and measurement problems are formidable. Military effectiveness is an elusive concept; the definition of an effective military organization is far from precise and the definition of a good soldier even less so. Imponderables such as discipline, leadership, training, societal influence, and group relationships all bear on efficiency.

In this chapter we examine three determinants of military effectiveness—individual capabilities, group performance, and the perceptions of foreign countries—and discuss the implications of each.

Individual Capabilities

Job performance depends on such characteristics as mental ability, level of education, job aptitude, physical condition, experience, motivation, adaptability to change, and ability to get along with co-workers.

---

1. Alvin J. Schexnider and John Sibley Butler, "Race and the All-Volunteer System: A Reply to Janowitz and Moskos," *Armed Forces and Society,* vol. 2 (Spring 1976), pp. 423–24.

All are to some extent interrelated, and their relative importance varies by type of job and experience level within a given occupation. But gauging how well individuals perform their jobs is a difficult task, especially in the skills—such as combat infantryman—for which standards of quantitative measurement in a peacetime environment are not particularly well developed. The relative capabilities of black and white soldiers have long been a debatable subject. As the discussion in chapter 2 illustrated, assessments are likely to differ when the origins, experiences, and philosophies of those who undertake them differ. Thus any comparison of individual capabilities must be drawn with caution.

Contemporary Indicators of Individual Performance

The military services assess individual capabilities in several ways. They have elaborate performance-report systems in which their members are periodically evaluated by their supervisors. These ratings play a prominent role in the promotion process, helping to determine how rapidly individuals will move up through the ranks. The services also use a variety of tests. For example, a skill qualification test administered to Army personnel is designed to evaluate a soldier's ability to perform up to thirty-five critical tasks specific to an occupational specialty and a particular duty position. Other criteria for evaluating performance, especially of junior members, include the degree of success in training courses, disciplinary records, and the ability to complete the first enlistment period. How appropriate these indicators of performance are, however, is questioned—in effect, the debate about the present state of the nation's military forces centers on them. Critics of the status quo link what they perceive as a diminished U.S. military capability to deterioration in performance; some who defend the all-volunteer force argue that the indicators do not adequately measure a soldier's effectiveness, and the extent to which they are correlated with a soldier's ability to perform on the job remains unknown.

These differences of opinion are difficult to reconcile since the evidence either way is less than conclusive. Nonetheless these performance indicators will continue to be used until more appropriate measures are found. This is important since the armed forces' standards for enlistment are rationalized largely by their validity as predictors of individual performance, which these indicators are purported to measure.

Predictors of Individual Performance

What characteristics foretell how well an individual will perform in a military setting? This puzzling question is made more difficult by the performance measurement problem described above and by the obvious fact that different jobs require different qualities. Adjectives such as "capable," "motivated," "productive," "suitable," "competent," "disciplined," and "adaptable" are all used to describe the desired attributes of the military recruit. But because of the difficulty of constructing individual profiles and predictors of performance and because of the myriad occupations in the armed forces, manpower "quality" is most often described in the easy-to-measure terms of educational level and standardized test score attainment; the *choice* recruit, according to the services, has a high school diploma and an above-average aptitude for military skills.

EDUCATIONAL LEVEL. The armed forces place a high premium on the completion of high school, not so much because of its relationship to mental achievement—although that is important for some technical courses—but because of its implications for general adaptability to the military environment. "The significance of a high school diploma," notes one scholar, "is that it reflects the graduate's motivation and ability to accomplish one of society's important goals, to complete one program before beginning another or, in other words, to stick to a project even if the going is difficult."[2]

The predictive validity of educational levels has also been pushed beyond disciplinary behavior:

In addition to overall military competence, educational achievement has been found to be related to such specific effectiveness criteria as performance in combat and in various military occupational settings. Using superior officer ratings of combat performance . . . marines with the highest combat ratings were more likely than not to be high school graduates.[3]

It is generally true that high school graduates are more likely to complete their term of enlistment; in fact, dropouts fail to complete their first three years of service at about twice the rate of those who have

---

2. Anne Hoiberg, "Meeting Personnel Needs," *Society*, vol. 18 (March–April 1981), p. 39.
3. Anne Hoiberg, "Military Staying Power," in Sam C. Sarkesian, ed., *Combat Effectiveness: Cohesion, Stress, and the Volunteer Military* (Sage, 1980), vol. 9, p. 215.

earned a diploma.[4] But a study of the conditions of separations and discharges of soldiers who had entered the Army between 1972 and 1975 yielded a surprising result: among black youths, high school graduates received proportionately more separations than their less educated peers for personality disorders, in lieu of courts-martial, and for expediency (poor attitude, lack of motivation, and the like).[5] The causes of this contradiction have not been investigated.

ENTRY TEST SCORES. Military aptitude standards play an important part in the personnel screening, classification, and assignment process. The services use mental aptitude tests and level of education as the principal predictors of trainability. A standard test, the Armed Services Vocational Aptitude Battery (ASVAB), now administered to all prospective volunteers is composed of ten subtests: general science, arithmetic reasoning, word knowledge, paragraph comprehension, numerical operations, coding speed, mathematics knowledge, electronics information, mechanical comprehension, and auto and shop information.[6] A composite, which includes word knowledge, arithmetic reasoning, paragraph comprehension, and numerical operations subtests, yields a single index of general trainability, known as Armed Forces Qualification Test (AFQT) scores. On the basis of these test scores, examinees are divided into the following groups representing the range from very high military aptitude (category I) to very low military aptitude (category V):

AFQT category	Percentile score	Net raw score (105 possible)
I	93–100	102–105
II	65–92	84–101
III	31–64	65–83
IV	10–30	38–64
V	9 and below	37 and below

4. *America's Volunteers: A Report on the All-Volunteer Armed Forces* (Office of the Assistant Secretary of Defense for Manpower, Reserve Affairs, and Logistics, 1978), p. 68. See also appendix table B-6.

5. Peter G. Nordlie and others, *A Study of Racial Factors in the Army's Justice and Discharge Systems*, HSR-RR-79/18-Hr, prepared for the DAPE-HRR, Department of the Army (McLean, Va.: Human Sciences Research, Inc., 1979), vol. 2, p. 42.

6. For a brief description of the version of the test battery introduced in 1980, see William H. Sims and Ann R. Truss, "Normalization of the Armed Services Vocational Aptitude Battery (ASVAB) Forms 8, 9, and 10 Using a Sample of Service Recruits" (Arlington, Va.: Center for Naval Analyses, December 1980).

The test scores are used principally to differentiate between categories I and II (above average), III (average), and IV and V (below average). Category III is sometimes divided into two groups, category IIIA encompassing those who score between the fiftieth and sixty-fourth percentiles and category IIIB those who score between the thirty-first and forty-ninth percentiles. Entrants scoring below the thirty-first percentile are considered by the services to require more training and present greater disciplinary problems than those in the higher groups, and those scoring below the tenth percentile are disqualified.[7]

Standardized tests have been administered by the armed forces since World War I. They were used initially to identify the mentally unfit and only more recently to assess vocational aptitude. The services place great stock in the tests, largely because research has shown that individuals with higher aptitude scores are more trainable; that is, they tend to complete skill training courses at a higher rate than low scorers, they complete them sooner, they get higher grades, and they retain the information longer.[8] Also, research has verified a close connection between entry aptitude and skill qualification test scores and rates of

---

7. The percentile ranges of these categories are still defined by the distribution of scores attained by male military personnel (enlisted and officer) during World War II. The categories were set up at that time for "administrative simplicity," according to the following rationale: "In general, the top third, those in Classes I and II, were able to cope with any training program, even the most difficult. Those in the middle third, Class III, could meet most training requirements, other than those that required ability to deal with mathematics and other abstract forms of reasoning. Men in Class IV were expected to absorb basic training and even some of the simpler types of advanced training, but this was not true about the more than 750,000 recruits in Class V. Men in the lowest class were definitely handicapped and could be expected to meet minimum performance standards only if they were given special training, such as a preliminary course to prepare them for basic training, or a prolongation of basic training itself." Eli Ginzberg and others, *The Ineffective Soldier: Lessons for Management and the Nation*, vol. 1: *The Lost Divisions* (Columbia University Press, 1959), p. 45.

8. A strong correlation between scores on entry tests and scores on skill qualification tests (SQT) has also been documented. A survey of 24,000 soldiers found the following relationship:

AFQT category	Average SQT score
I	80
II	73
IIIA	68
IIIB	63
IV	59

Based on data provided by the Department of Defense, Defense Manpower Data Center.

promotion.⁹ But here again, the curious inverse relationship that was found between level of education and honorable separations among black soldiers also applies to entry test scores: "with respect to type of discharge . . . being in a higher mental category is not an advantage for blacks."¹⁰

Validity of Entry Standards

The entry standards applied by the armed forces, particularly the standardized entry tests, raise several issues besides that about the link between test scores and job performance per se. Whether the tests measure aptitude, achievement, or intelligence, whether they are truly divorced from the influences of education and environment, and whether they are free of cultural bias are among the many questions being asked. The tests are also criticized on the grounds that they fail to measure such attributes as idealism, stamina, persistence, and creativity, which many observers consider to be just as important as cognitive skills in contemporary American society. The central charge, however, is that standardized tests do not measure the same dimensions of achievement across different groups.

STANDARDIZED TESTS AND MILITARY PERFORMANCE. Many critics show a growing tendency to measure military capability, and indeed to assess the concept of voluntary recruitment, by the scores of recruits on entry tests. In response to a seemingly growing intolerance in Congress of the upward trend in the proportion of volunteers scoring below the thirty-first percentile, an ad hoc study group in the Army secretariat examined the validity of standardized tests and reported:

The Army cautiously states that results of the AFQT indicate, at best, trainability. The evidence we have gathered, however, suggests that the test has been so

---

9. The connection between entry test scores and the rate at which Army personnel rise through the ranks is shown below by the average number of months served at the time of promotion to grade E-5 (based on data provided by the Defense Manpower Data Center):

AFQT category	1972	1974	1976	1978	First half, 1980
I	23.6	32.0	34.7	38.7	40.1
II	26.4	33.9	37.2	39.9	41.7
IIIA	31.3	36.6	39.4	42.0	44.9
IIIB	35.3	39.1	40.3	42.9	45.9
IV	34.6	37.6	41.9	45.5	44.3

10. Nordlie and others, *Study of Racial Factors*, vol. 2, p. 46.

misrepresented over time, and the predictions derived from Mental Category results so overstated, that the future utility of the AFQT is in some doubt.[11]

The study group argued that "linking Mental Category with job performance is not only inaccurate but against the best interests of manpower management in the Army—which requires finding the soldier who can do the job."[12] To this end, they suggested that the Army convert from a norm-referenced system to a criterion-referenced system. Whereas the former compares one candidate with others, the latter would compare the candidate with a describable, constant standard of performance.[13]

ERRORS OF REJECTION. The use of aptitude scores to eliminate entire groups of applicants can be criticized on the ground that many individuals are rejected who if allowed to enlist would have succeeded. Estimates of the magnitude of this "error of rejection" have been made in several studies.

Under Project One Hundred Thousand the armed services waived certain aptitude requirements so that disadvantaged members of society might share in the opportunities provided by military service. Almost a quarter of a million men had entered the armed forces under this program by 1969; 45 percent of them were high school graduates, the median AFQT score was 13.6, and the median reading ability was at the sixth grade level.[14] The motivations, the implementation, and the results of that experiment are debatable, but the conclusions contained in the Pentagon's analysis of the project are quite clear:

As could be expected, the men brought in under reduced mental standards do not perform as well as a cross section of men with higher test scores and educational abilities. This is true on all measures—training attrition, promotions, supervisory ratings, disciplinary record, and attrition from service. The differences are not large and we feel they are acceptable when balanced against the military and social goals of the program.[15]

---

11. "An Examination of the Use of the Armed Forces Qualification Test (AFQT) as a Screen and a Measure of Quality," Report to the Secretary of the Army and the Chief of Staff (Department of the Army, July 1980), p. iii. This study, known as the Lister Report after the Army's General Counsel Sara E. Lister, was done during Secretary of the Army Clifford Alexander's term. It is no secret that the military leadership was at odds with the secretary on many issues, including this one.

12. Ibid., p. III-11.

13. Ibid., p. IV-6. Criterion-referenced systems are becoming more widely discussed in the literature. See especially W. James Popham, *Criterion-Referenced Measurement* (Prentice-Hall, 1978).

14. *Project One Hundred Thousand: Characteristics and Performance of "New Standards" Men* (Office of the Assistant Secretary of Defense for Manpower and Reserve Affairs, 1969), pp. x, xiii.

15. Ibid., p. xxviii.

A recent assessment of the capabilities of those not normally considered eligible for military service was provided by an analysis of Army enlisted personnel who, because of errors in converting raw scores into percentile scores, were permitted to volunteer between January 1976 and September 1980. Had the test been properly calibrated, these soldiers would have been barred from enlistment under the Army standards in effect during the period.

Compared to their contemporaries who would have been eligible anyway, these soldiers, like their Project One Hundred Thousand predecessors, had lower scores on skill qualification tests and a higher first-term attrition rate. Yet the study pointed out that "it is important to recognize that the majority of [these soldiers] were successful."[16] The performance of this group (called "potentially ineligible") contrasted with the group "all others" as shown below:[17]

Performance measure	Potentially ineligible	All others
Skill training course attrition rate (percent)	9	7
Skill qualification test		
Average score	62	68
Percent passing	58	70
First-term attrition rate (percent)	48	35
First-term completers eligible to reenlist (percent)	72	75
First-term retention rate (percent)	23	25
Pay grade achieved after 3 years (percent)		
E-4 and E-5	76	85
E-5	4	9

The measurements were taken in thirty-four training courses covering a wide range of skills including combat, equipment maintenance, communications, and administration. While some were more difficult than others, most were not academically demanding. Slower learners were given more attention and more time to complete the course.

Neither analysis quantified the incremental costs associated with converting "substandard" individuals into effective performers. The armed forces may be systematically excluding capable and available

16. I. M. Greenberg, "Final Report: Mental Standards for Enlistment Performance of Army Personnel Related to AFQT/ASVAB Scores," in "Implementation of New Armed Services Vocational Aptitude Battery and Actions to Improve the Enlistment Standards Process" (Office of the Assistant Secretary of Defense for Manpower, Reserve Affairs, and Logistics, December 1980), pp. 1–69.

17. Ibid., p. 65.

candidates in what may prove to be an expensive search for those with the ability to score higher on military aptitude tests.

TEST BIAS. The Army's critique of standardized tests also raised the question of racial bias in the test instruments.[18] Specifically, the report indicated that certain words included in the word knowledge subtest "will disproportionately reflect a cultural background typical of the majority male population of white test-takers."[19] But there have been relatively few charges of cultural bias in military testing, which is surprising in view of the growing criticism of standardized tests in the nonmilitary sector.[20] In fact, there is no record of litigation brought against the Department of Defense or any of the services on test bias grounds.[21] In any event, a good deal of research has been done over the years, particularly by the Air Force, to test the validity of military entry tests for both minorities and females. In general, these analyses have indicated that the tests meet appropriate federal standards and that equations for predicting training performance (in terms of final school grade) are essentially the same for whites and minorities and for males and females. Where differences were observed, the predictions for the training school grades of minorities were too optimistic; in other words, the test predicted minority examinees would do better in training than they actually did.[22] Finally, in a technical assessment of the test battery, it was found that responses were "free from major defects such as high levels of guessing or carelessness, inappropriate levels of difficulty, cultural test-question bias, and inconsistencies in test administration procedures"; the investigators therefore concluded that the test results "provide a sound basis for the estimation of population attributes such

18. Bias in testing is a complex, highly technical question on which it is difficult to obtain agreement, even about its meaning. For an overview of the literature, see Mark J. Eitelberg, "Subpopulation Differences in Performance on Tests of Mental Ability: Historical Review and Annotated Bibliography," Technical Memorandum 81-3 (Directorate for Accession Policy, Office of the Secretary of Defense, August 1981).

19. "Examination of the Use of the Armed Forces Qualification Test," p. III-30.

20. For a general discussion of cultural bias in standardized testing, see Robert L. Green and Robert J. Griffore, "The Impact of Standardized Testing on Minority Students," *Journal of Negro Education*, vol. 49 (Summer 1980), pp. 238–62.

21. Letter from the Office of General Counsel, Department of Defense, May 6, 1981.

22. Nancy Guinn, Ernest C. Tupes, and William E. Allen, "Cultural Subgroup Differences in the Relationships Between Air Force Aptitude Composites and Training Criteria" (Lackland Air Force Base, Texas: Air Force Human Resources Laboratory, September 1970). See also Lonnie D. Valentine, "Prediction of Air Force Technical Training Success from ASVAB and Educational Background" (Lackland AFB: Air Force Human Resources Laboratory, May 1977).

as means, medians, and percentile points, for the youth population as a whole and for subpopulations defined by age, sex, and race/ethnicity."[23]

The issue is far from settled. On one side, the ASVAB, imperfect though it may be, remains the best instrument available for selection purposes, and the burden of proof rests on those who argue that the test is flawed to the point of being discriminatory. The other side holds that aptitude testing as now constituted does discriminate against blacks and other minorities and that the armed services should prove that the test is a valid predictor of military performance.[24] In any event, current standardized entry tests will probably continue to be used to screen potential military recruits. Indeed, by legislating maximum limits on the proportions of recruits who score below the thirty-first percentile, the Ninety-sixth Congress effectively institutionalized the standardized test as a measure of individual capability.[25]

Racial Comparisons

The racial implications of using educational and entry test standards are well recognized: blacks are less likely than whites to qualify for military service. Despite the impressive progress of the past four decades, relatively fewer young blacks than whites complete high school (see table 5-1). A wider disparity is found between the standardized test scores attained by each group (see table 5-2): about 30 percent of military-age blacks fall into category V (and thus are barred from entering the armed forces) but only about 4 percent of comparably aged white youths. Of those whose scores would place them in the top four categories, the average black youth is more likely to be in the lowest acceptable category (IV) and hence less likely to qualify for most technical training programs.[26]

23. R. Darrell Bock and Robert J. Mislevy, "The Profile of American Youth: Data Quality Analysis of the Armed Services Vocational Aptitude Battery" (University of Chicago, National Opinion Research Center, August 1981), p. 51.

24. It has been suggested that while these critics might have the interests of minorities at heart, their approach could be counterproductive. In an editorial opinion, the *Washington Post* (February 10, 1982) concluded: "Those who would throw out standardized tests because of poor minority group scores . . . are doing no one—especially their theoretical beneficiaries—a favor. On the contrary, they are undermining the credible argument for a stronger effort to reconcile these disparities—and thereby doing those minorities and the society as a whole great harm."

25. Department of Defense, Authorization Act, 1981, Conference Report, Title III, sec. 302.

26. Two caveats are appropriate here. First, statistical comparisons necessarily depend on averages of population subgroups, which often obscure the fact that many members of

Table 5-1. Males Aged 18 to 21 Who Completed High School, by Race, Selected Years, 1940–80
Percent

Race	1940	1950	1960	1970	1980
White	40	49	56	68	78
Black	11	18	33	49	60

Sources: 1940, *Sixteenth Census*, vol. 4, pt. 1 (Government Printing Office, 1943), pp, 43–44; 1950, *Census of Population, 1950*, vol. 2, pt. 1 (GPO, 1953), p. 225; 1960, *Census of the Population, 1960*, vol. 1, pt. 1 (GPO, 1964), table 172, p. 399; 1970, *Census of Population, 1970*, vol. 1, pt. 1, sec. 2, table 198, p. 621; 1980, *Current Population Reports*, Series P-20, no. 362, "School Enrollment: Social and Economic Characteristics of Students: October 1980 (Advance Report)" (GPO, 1981).

As pointed out above, in addition to the single index of general aptitude (the AFQT), the test battery includes a variety of subtests that are combined to form composites designed to predict training success for clusters of occupations. The services use aptitude subtest results in different combinations tailored to their respective needs. To qualify for enlistment in the Army in 1981, for instance, high school graduates were required not only to score at the sixteenth percentile or higher on the AFQT, but also to attain a score of 85 or above on at least one aptitude composite. Non-high school graduates, on the other hand, had to score at the thirty-first percentile or higher on the AFQT and also attain a score of 85 or higher in two aptitude areas. Minimum composite scores were established for entry into various occupations as well; for example, to enter training as a carpenter, mason, or electrician, an enlistee had to attain a score of at least 85 in the general maintenance aptitude composite, and to enter the law enforcement field, a volunteer had to attain a minimum score of 95 on the skilled technical composite. The formulas and cutoff scores vary by service but the principles are similar.

To illustrate, the occupational areas for which composites are calculated by the Army and the subtests included in each are shown in table 5-3. Comparison of the performance by black and white youths on the various subtests reveals consistent differences; in all subtests whites exceed blacks by about one to one and a half standard deviations of the total population score distributions (see table 5-4).[27] These results are

---

the group may score above or below the average for their particular group or, for that matter, any other group. Second, we have not attempted to analyze the possible factors that account for subgroup differences. For a discussion of this issue, including its environmental and genetic dimensions, see R. Darrell Bock and Elsie G. J. Moore, "Advantage and Disadvantage: Vocational Prospects of American Young People" (National Opinion Research Center, forthcoming).

27. These differences are substantial. Roughly speaking, a spread of one standard deviation means that a cutoff score that included the top 50 percent of the whites would

Table 5-2. Estimated Percentage Distribution of Males' Entry Test Scores Referenced to 1944 Standard, by Race, Selected Years, 1941–80[a]

AFQT category	1944 reference population	1941–46[b] White	1941–46[b] Black	1960 White	1960 Black	1971–72 White	1971–72 Black	1980 White	1980 Black
I	8	6	*	10	*	8	*	6	*
II	28	29	3	29	4	38	5	41	7
III	34	33	13	38	24	37	27	31	20
IV	21	26	48	18	41	12	40	19	43
V	9	6	36	5	31	4	29	4	30

Sources: 1941–46 distribution from H. S. Milton, ed., *The Utilization of Negro Manpower in the Army*, Report ORO-R-11 (Chevy Chase, Md.: Operations Research Office, Johns Hopkins University, 1955), p. 10; 1960, Bernard D. Karpinos, "The Mental Qualification of American Youths for Military Service and Its Relationship to Educational Attainment," unpublished paper, 1966; 1971–72, unpublished tabulation by Bernard D. Karpinos; 1980, Department of Defense, "Profile of American Youth: 1980 Nationwide Administration of the Armed Services Vocational Aptitude Battery" (Office of the Assistant Secretary of Defense for Manpower, Reserve Affairs, and Logistics, March 1982). Figures are rounded.

* Less than 0.5 percent.

a. The distributions shown for each time period are for a somewhat different population: the 1944 reference population distribution is based on the test scores attained by every member of the military (enlisted and officer) who was on active duty in December 31, 1944; the 1941–46 estimate on scores of Army inductees; the 1960 and 1971–72 estimates on an analysis of tests administered to all preinductees (those ordered to report for induction examination); and 1980 data on tests administered to a cross section of American youths aged eighteen to twenty-three.

b. These data overstate the scores that would have been attained by the general population since roughly one out of every ten whites and one out of every three blacks were rejected on mental and educational grounds (those who could not read, write, or do sums at the fourth grade level). See Eli Ginzberg and others, *The Ineffective Soldier: Lessons for Management and the Nation*, vol. 1: *The Lost Divisions* (Columbia University Press, 1959), p. 121. If those rejected are counted as category Vs, the percentages are:

Category	White	Black
I	5	*
II	26	2
III	30	9
IV	24	32
V	15	57

Table 5-3. Aptitude Subtest Components, by Army Occupational Area

Occupational area	Subtest components
Maintenance	General science, auto and shop information, mathematics knowledge, electronics information
Electronics	General science, arithmetic reasoning, mathematics knowledge, electronics information
Clerical	Numerical operations, coding speed, word knowledge
Mechanical maintenance	Numerical operations, auto and shop information, mechanical comprehension, electronics information
Surveillance and communications	Numerical operations, coding speed, auto and shop information, mechanical comprehension
Combat	Arithmetic reasoning, coding speed, auto and shop information, mechanical comprehension
Field artillery	Arithmetic reasoning, coding speed, mathematics knowledge, mechanical comprehension
Operator and food service	Numerical operations, auto and shop information, mechanical comprehension, word knowledge
Skilled technical	General science, mathematics knowledge, mechanical comprehension, word knowledge

Source: Department of Defense, "Manual for Administration: Armed Services Vocational Aptitude Battery, Forms 8, 9 and 10," October 1979.

not surprising since the opportunities for blacks to have contact with and participate in the "majority culture from which the vocational test materials are drawn" have been so limited.[28]

Since under present standards technical courses require high aptitude scores in many of the subtests in which the largest racial differences are found, fewer blacks are likely to qualify for skill training. This is illustrated in table 5-5, which compares the mean scores of white and black youths in four occupational areas used by the Air Force in its classification process. Substantial differences exist in each occupational area, but blacks are at a particular disadvantage in qualifying for training that requires relatively higher composite scores in the mechanical and electronics occupations.[29]

The net effect on racial groups of the aptitude standards discussed above is summarized in table 5-6, which shows the estimated proportions of white and black males eighteen to twenty-three years old who would have qualified for enlistment in the military services under the eligibility standards in effect in fiscal 1981. It is clear, first of all, that enlistment "selectivity" varies from service to service. It is also clear that the difference in eligibility rates for whites and blacks in every service is quite substantial. For example, approximately four out of five white

---

include only 16 percent of the blacks. This is based on the assumption that the scores of blacks and whites are distributed normally and that the standard deviations of each distribution are the same.

28. R. Darrell Bock and Elsie G. J. Moore, "The Profile of American Youth: Demographic Influences on ASVAB Test Performance," Executive Summary (National Opinion Research Center, December 1981), p. 17.

29. A high level of correlation (.85) has been found between ASVAB "general" composite and Adult Basic Learning Examination (ABLE) scores. The conversion of general composite scores to ABLE provides the following estimates of reading grade levels for the national youth population as of 1980:

	Males				Females			
	Mean		Standard deviation		Mean		Standard deviation	
Age group	White	Black	White	Black	White	Black	White	Black
18–19	9.6	6.8	2.3	2.1	9.5	7.0	2.1	1.9
20–21	10.0	7.0	2.3	2.2	9.8	6.9	2.1	2.1
22–23	10.6	7.2	2.0	2.4	10.0	7.3	2.1	2.0
18–23	10.1	7.0	2.2	2.3	9.8	7.1	2.1	2.0

Based on data in "Profile of American Youth: 1980 Nationwide Administration of the Armed Services Vocational Aptitude Battery" (Office of the Assistant Secretary of Defense for Manpower, Reserve Affairs, and Logistics, March 1981), p. 82.

Table 5-4. Comparative Means and Standard Deviations of Raw Scores on Armed Services Vocational Aptitude Battery, 18- to 23-Year-Old Males, by Race, 1980

Subtest	Number of questions	Mean			Standard deviation		
		White	Black	Total[a]	White	Black	Total[a]
General science	25	18.0	11.7	16.8	4.6	4.8	5.2
Arithmetic reasoning	30	20.5	12.2	19.0	7.1	5.7	7.5
Word knowledge	35	27.9	18.4	26.2	6.7	8.7	7.9
Paragraph comprehension	15	11.3	7.9	10.7	3.1	3.7	3.5
Numerical operations	50	35.3	25.1	33.5	10.2	11.4	11.1
Coding speed	84	45.4	30.7	42.9	14.8	14.9	15.7
Auto and shop information	25	18.6	10.8	17.2	4.6	4.6	5.5
Mathematics knowledge	25	15.0	9.4	14.0	6.5	5.0	6.6
Mechanical comprehension	25	17.4	10.4	16.2	4.8	4.3	5.4
Electronics information	20	14.0	8.7	13.1	3.7	3.9	4.2

Source: Derived from special tabulations provided by the Office of the Assistant Secretary of Defense for Manpower, Reserve Affairs, and Logistics. Figures represent results of administration of the ASVAB in 1980 to a national probability sample that included over 4,500 American males aged eighteen to twenty-three.
a. Total includes all racial and ethnic groups.

males but fewer than half of all black males could have qualified for enlistment in the Army during fiscal 1981. And the disparity between whites and blacks is even wider in the other services. About three out of ten white young men, for instance, could not qualify for entry into the Air Force; in sharp contrast, almost four out of five blacks probably would be rejected by this service.

Among black high school dropouts, the prospects for serving in the nation's military were even more bleak. Blacks without a high school

Table 5-5. Comparative Means and Standard Deviations of Percentile Scores of 18- to 23-Year-Olds on Selected Aptitude Composites, by Race and Sex, 1980

	Males				Females			
	Mean		Standard deviation		Mean		Standard deviation	
Composite[a]	White	Black	White	Black	White	Black	White	Black
Mechanical	56.9	20.3	22.7	16.7	28.8	11.9	14.8	8.2
Administrative	47.5	21.0	23.3	18.8	56.1	29.0	23.3	20.6
General	56.6	24.0	23.8	20.5	52.8	23.5	23.3	17.9
Electronics	57.7	23.9	23.0	20.3	45.7	19.5	22.9	16.2

Source: Department of Defense, "Profile of American Youth," pp. 86–89.
a. Scores are calculated with formulas that include the scores on the following subtests: *mechanical:* auto and shop information, mechanical comprehension, and general science; *administrative:* word knowledge, coding speed, numerical operations, and paragraph comprehension; *general:* word knowledge, arithmetic reasoning, and paragraph comprehension; *electronics:* arithmetic reasoning, electronics information, general science, and mathematics knowledge.

Table 5-6. Males Aged 18 to 23 Eligible for Enlistment Based on Fiscal 1981 Aptitude Test Standards, by Level of Education and Race
Percent

Level of education and race	Army	Navy	Marine Corps	Air Force
High school graduates				
White	96.1	95.5	92.1	88.1
Black	66.2	63.7	52.1	34.9
Non-high school graduates[a]				
White	47.0	40.8	43.8	18.4
Black	11.6	9.3	8.4	1.7
All males				
White	84.3	82.3	80.5	71.3
Black	43.8	41.4	34.2	21.3

Sources: Derived from special tabulations provided by the Office of the Assistant Secretary of Defense for Manpower, Reserve Affairs, and Logistics; Defense Manpower Data Center; and Department of Defense, "Profile of American Youth."
a. Includes males who have passed General Educational Development high school equivalent.

diploma in fiscal 1981 had little or no likelihood of being able to meet the minimum enlistment criteria established by the armed forces; fewer than 12 percent could pass Army standards and fewer than 2 percent could meet those of the Air Force. The situation was better for black high school graduates; one of every three could have qualified under Air Force aptitude criteria and two of every three could have scored above minimums on the Army tests. In each case, however, blacks were far less likely than whites to qualify for the armed forces.

Group Performance

Thus far the discussion of comparative capabilities of whites and blacks in the military has centered on the attributes of the individual. In discussing the effectiveness of military units, however, the importance of group relationships cannot be overlooked. Does the presence of a disproportionate number of members of minority groups adversely affect the abilities of military units to perform their mission? Some believe that racial tension adversely affects unit cohesion. Others worry that growth in minority representation leads to a decline in the quality of white volunteers and hence to an overall decline in military capability. Still others fear that the capabilities of military units could be compromised in situations where the deployment of forces might test the allegiance of minority members.

Extent of Integration and Its Effect on Group Performance

Previous studies of racial integration in the armed forces were based on a military in which overall black participation did not exceed the proportion of blacks in the total population; in effect, the military was a predominantly "white" institution.

As the proportion of blacks in the armed forces began to grow, so did concern about the possible impact on group relations. As early as 1974 Congress directed the Defense Manpower Commission, a group set up to examine a range of manpower issues, to give special consideration to "the implications for the ability of the armed forces to fulfill their mission as a result of the change in the socioeconomic composition of military enlistees."[30]

In its interim report in 1975 the commission said:

Whereas the draft in essence gave the military a distribution system which caused the Services to more nearly reflect a cross section of American youth, the All Volunteer Force, on the other hand, leaves the socio-economic and minority group balance more to chance. The result has been some pronounced aberrations in the distribution of military personnel which, although solvable, nevertheless create problems.[31]

In its final report in 1976, however, the commission "found no evidence" that "the socioeconomic composition of a force affects its performance."[32] This conclusion was largely based on an investigation by the commission staff, which administered a survey to 154 commanders of military units that included the question: "How has the increase of blacks and Spanish-speaking Americans affected the ability of your unit to carry out its mission?" Forty-four percent perceived no increase; of those who did, 88 percent reported that the increase had no effect on

---

30. Public Law 93-155, sec. 702(7).
31. Defense Manpower Commission, "Interim Report to the President and Congress," May 16, 1975, p. 14.
32. Defense Manpower Commission, *Defense Manpower: The Keystone of National Security*, Report to the President and Congress (Government Printing Office, 1976), p. 156. What caused the change in position is unclear. The composition of the commission had been criticized by Eddie Williams, the president of the Joint Center for Political Studies: "It is discouraging to note . . . that not one of the seven commissioners is black and that there is only one permanent black professional [on the staff]. . . . This would appear to be a case of underrepresentation." *Focus* (monthly newsletter of the Joint Center for Political Studies, June 1975), p. 2.

unit capabilities, 5 percent indicated that the increase had a positive effect, and the remaining 7 percent reported a negative influence.[33] The commission staff also reported "a higher degree of interest in this subject in the academic, political and higher military levels than at the level of the operating military unit." Finally, the report hedged: "as with *any* change (i.e. people, time, training, etc.) a true evaluation can only be arrived at after a unit is committed to actual combat."[34]

Yet by the end of the decade race relations in the armed forces appeared to have taken a turn for the worse. The relative racial tranquility that accompanied the end of the Vietnam era and the transition to voluntary recruitment gave way to what was described by a senior Pentagon official as "a new racism" in the military, ostensibly a backlash to affirmative action programs.[35] Some social scientists were reporting similar phenomena in American society. According to the results of one survey, the growth of racial liberalism among whites, which was so evident in the 1960s, abated and in some respects was reversed, particularly in attitudes toward housing and "intrusion." A mood of white intransigence was reflected as more and more whites expressed the opinion that "Negroes should not push themselves where they are not wanted."[36] Other polls, however, reported a continued lessening of racial tension, as measured by the trend in favorable responses of whites to integrated schooling, marriage, and "intrusion."[37]

At the same time, many blacks were apparently becoming increasingly disenchanted as their heightened expectations—the legacy of the civil

33. Kenneth J. Coffey and others, "The Impact of Socio-Economic Composition in the All Volunteer Force," *Defense Manpower Commission Staff Studies and Supporting Papers,* vol. 3: *Military Recruitment and Accessions and the Future of the All Volunteer Force* (GPO, 1976), p. E-66. Survey results varied among the services. Sixty-four percent of the Army commanders discerned an increase in minorities; of these, 83 percent reported no effect on unit capability, 3 percent a positive effect, and 14 percent a negative effect.

34. Ibid., p. E-51.

35. Comments attributed to M. Kathleen Carpenter, deputy assistant secretary of defense for equal opportunity. "Pentagon Official Reports 'a New Racism' in Military," *New York Times,* July 24, 1979. This view was shared by Sharon Lord, Carpenter's successor in the Reagan administration: "There is a great deal of evidence that racism continues to exist in very subtle forms." *Air Force Times,* January 11, 1982, p. 12.

36. John G. Condran, "Changes in White Attitudes Toward Blacks: 1963–1977," *Public Opinion Quarterly,* vol. 43 (Winter 1979), pp. 463–76; quotation is from p. 464.

37. "Opinion Roundup," *Public Opinion,* October–November 1980, p. 28. See also "A New Racial Poll," *Newsweek,* February 26, 1979, p. 48, for the results of a 1978 Harris survey, which concluded that, in comparison to 1963, whites had become far more tolerant of integration, less given to racial stereotyping, and ready to accept wide-ranging affirmative action programs.

rights movement—continued to be largely unfulfilled, with little promise of improvement. Particularly evident was the new conception of the black man in America, described by Elijah Anderson in 1980:

The newly emerging picture, be it widely applicable to blacks or not, is that of an assertive, struggling young person who will not "take the shit" his forefathers did. This new black person is especially sensitive to what could be construed as racial prejudice, slight, and discrimination. . . . Add to this new image of militancy the growing problem of youth unemployment and a stereotype of black youth as being primarily responsible for urban street crime, street gang activity, and general incivility, and one is faced with the specter of a nearly "unemployable" person.[38]

And restlessness is evident even among better-situated blacks who had reaped rewards from the civil rights movement, as described by Roger Wilkins:

Young professionals who were of college age or just a little older during the black-consciousness movement in the late Sixties came to adulthood feeling indifferent about white people. Because they work in a more integrated society than any blacks in the history of the country, they know more about white people than any earlier generation. Across a broad spectrum, they seem to have concluded that working relationships with whites are enough. In their leisure time, they generally prefer the company of people who share their own deep and troubling American experiences and insights.

This black attitude stems in part from the fact that life in the newly integrated workplace is fraught with more racial friction than most white Americans understand or acknowledge. When it comes to socializing, blacks don't want to face the same stressful situations they encounter on the job.[39]

Whether or not race relations in America are deteriorating, concern has been expressed about tension that might arise in military units with adverse effects on unit cohesion, on the one hand, and the prospect that the armed forces might reach a "tipping point" where qualified white youths were deterred from volunteering for military service, on the other.

SOCIAL NETWORKS AND UNIT COHESION. In the past, the influence of social networks in military organizations, and in combat settings in particular, was the subject of extensive social research.[40] Wartime

38. Elijah Anderson, "Some Observations on Black Youth Employment," in Bernard E. Anderson and Isabel V. Sawhill, eds., *Youth Employment and Public Policy* (Prentice-Hall, 1980), p. 85.

39. Roger Wilkins, "The Widening Social Chasm," *Fortune*, March 9, 1981, p. 115.

40. See, for example, David G. Mandelbaum, *Soldier Groups and Negro Soldiers* (University of California Press, 1952); Samuel A. Stouffer and others, *The American Soldier: Adjustment During Army Life*, vol. 1 (Princeton University Press, 1949); Edward A. Shils, "Primary Groups in the American Army," in Robert K. Merton and Paul F.

experiences indicated that racial integration in the armed forces did not diminish U.S. military capabilities. Yet no recent systematic analysis has come to grips with a question left hanging by the earlier research: do the relative proportions in the racial mix in a military unit have any particular effect on cohesion and hence performance?

In the early 1950s, at the same time that sociologists were expounding the virtues of integrated groups in a military setting, a word of caution was added: "Integration may not have the same consequences as those heretofore recorded when instituted . . . in a unit which is predominantly Negro rather than predominantly white."[41]

In a major study of the use of blacks in the Army during the Korean War, Johns Hopkins researchers concluded: "The performance of a unit in combat or garrison is not adversely affected when integration is carried out under the usual circumstances in which Negroes are a minority. A maximum of 15-20 percent Negro personnel seems to be an effective interim working level."[42] The same study, however, suggested that there were no practical limits on integration. This was based on a regimental commander's ratings of the performance of twenty-eight training companies with varying proportions of blacks:[43]

Percentage of blacks in company	Rating		
	Above average	Average	Below average
1-5	2	4	0
6-9	5	5	0
10-35	4	2	2
63-92	2	2	0

Understandably, this line of research was not pursued during the 1950s and 1960s inasmuch as a racial mix proportionate to the population

---

Lazarsfeld, eds., *Continuities in Social Research: Studies in the Scope and Method of "The American Soldier"* (Free Press, 1950); Roger W. Little, "Buddy Relations and Combat Role Performance," in Morris Janowitz, ed., *The New Military* (Russell Sage Foundation, 1965); and Charles C. Moskos, Jr., *The American Enlisted Man: The Rank and File in Today's Military* (Russell Sage Foundation, 1970).

41. Mandelbaum, *Soldier Groups*, p. 132.

42. H. S. Milton, ed., *The Utilization of Negro Manpower in the United States Army*, Report ORO-R-11 (Chevy Chase, Md.: Operations Research Office, Johns Hopkins University, 1955), p. 5.

43. Ibid., p. 364. The two units that received "below average" ratings, one 34 percent black and the other 35 percent black, were reported by division officers to have a high proportion of members with physical limitations, which affected their performance.

was widely viewed as the maximum feasible upper limit of integration. And it is equally understandable that the question was not posed in the 1970s even as the mix changed sharply; to do so would have highlighted potential racial problems at the time the armed services were attempting to downplay their existence.[44]

But few would deny that the social networks in contemporary military organizations are somehow affected by the racial proportions of the unit. This does not necessarily imply that group cohesion deteriorates as the proportion of blacks increases; on the contrary, it could well turn out that an increase in the number of blacks—even beyond the 50 percent level—could improve race relations. This would happen, for example, if the morale of black soldiers was being adversely affected by perceived discrimination that they might associate with their minority status.

On the other hand, the possibility that a more nearly equal racial mix might solidify intrarace friendships and strengthen racial cliques cannot be ignored. And if the composition changed enough for the minority to become the majority, some of the same problems that were attributed to segregated units in World War II and Korea might crop up once again. These units were ineffective, by one account, because segregation was "a continuous reminder to the men that a low opinion of their Negro fellows is held by parts of the larger society."[45] To the extent, then, that blacks in the armed forces perceived a decline in whatever prestige they might associate with being "good enough" to be members of a predominantly white unit, cohesion and effectiveness could be adversely affected.[46]

Surveys of attitudes of military personnel about race relations, despite inherent limitations, provide some insight into the situation. In fiscal 1979, although a relatively small proportion of junior enlisted soldiers (grades E-1 through E-5) identified race problems as the greatest personnel problem in the unit, blacks were more inclined to hold that view, by roughly 2 to 1. To the question "Over the last six months, has the racial

44. The armed forces were not so reticent when it came to women's integration. In 1976 a test (called "MAX-WAC") was initiated by the Army to assess the effects of varying percentages of female soldiers assigned to representative types of units on the capability of a unit to perform its mission. See Department of the Army, "Outline Test Plan—Women Content in Units (MAX-WAC)," FO 048 (May 21, 1976).

45. Mandelbaum, *Soldier Groups,* p. 131.

46. Apparently, many blacks have the same negative opinions of blacks as white racists have. A 1981 *Washington Post*/ABC News Poll reported that 47 percent of black respondents felt that "most blacks don't have the motivation or will power to pull themselves out of poverty," and 23 percent indicated that "blacks would rather accept welfare than work for a living." See *Washington Post,* March 26, 1981.

situation in your unit improved or gotten worse?" about 15 percent of each racial group responded "worse." The proportion of blacks so responding in 1979 was roughly the same as in 1976; by contrast, the proportion of whites with that view had increased from a low of roughly 9 percent in 1976.[47]

When soldiers were asked in another survey how frequently soldiers of their own race "complained about better treatment being given to people of other races or ethnic groups in the armed forces," the responses varied by race and by the proportion of minorities in the unit (in percent):[48]

Proportion of minorities in respondent's unit	Frequency of observation					
	Very often or often		Sometimes		Seldom or never	
	White	Black	White	Black	White	Black
Most or more than half	31	44	28	29	41	27
About half	23	32	29	28	48	40
Some or a few	18	42	26	25	56	33

A similar pattern was found in responses to a question about the frequency with which soldiers of their own race "avoided doing things with people of other races or ethnic groups":

Proportion of minorities in respondent's unit	Frequency of observation					
	Very often or often		Sometimes		Seldom or never	
	White	Black	White	Black	White	Black
Most or more than half	27	17	24	20	49	63
About half	17	14	25	30	58	56
Some or a few	12	19	21	22	67	59

And finally, to a question about how often soldiers of their own race

47. *Equal Opportunity: Fourth Annual Assessment of Military Programs* (Office of the Deputy Chief of Staff for Personnel, Department of the Army, 1980), apps. 41 and 42.

48. Derived from data provided by the Defense Manpower Data Center. The responses were included in the "1978/79 Department of Defense Survey of Officers and Enlisted Personnel." For a discussion of the survey's methodology and results, see Zahava D. Doering and William P. Hutzler, *A Description of Officers and Enlisted Personnel in the U.S. Armed Forces: A Reference for Military Manpower Analysis*, R-2851-MRAL (Santa Monica: Rand Corp., March 1982).

"spoke badly or told racist jokes about people of other races or ethnic groups," the responses were distributed as follows:

Proportion of minorities in respondent's unit	Frequency of observation					
	Very often or often		Sometimes		Seldom or never	
	White	Black	White	Black	White	Black
Most or more than half	28	15	27	28	45	57
About half	21	17	30	22	49	61
Some or a few	21	25	27	22	52	53

The survey results show that most white and black soldiers "seldom or never" observe "avoidance" or racial slurs, regardless of the proportion of minorities in the unit. Black soldiers, on the other hand, report a generally higher incidence of complaints about unfair treatment; and this is especially the case in units that have very high or very low proportions of minorities. At the same time, the survey results suggest that the proportion of white soldiers who "very often or often" observe complaints, avoidance, and racial slurs in other whites is influenced by the composition of the unit—the larger the proportion of minorities, the greater the perceived tension in race relations. For black soldiers, the correlation between unit content and racial stress is less clear; from responses to the first question, at least, it appears that black soldiers believe the strain in race relations is minimized when the unit is composed of an equal mix of whites and blacks.

WHITE FLIGHT. Proponents of the "tipping point" theory of race relations argue that when the proportion of blacks in a neighborhood, a school, or even a city reaches a certain level, the number of whites begins to decline. The phenomenon is often attributed to whites' fear of black "hooliganism," vandalism, and an inevitable decline in property values. These notions remain fairly pervasive in contemporary American society, and it should come as no surprise that the theory has been extended to the nation's military as its minority content has increased.[49]

---

49. This is not to say that the issue has not been previously raised in a military context. A major finding in a study of blacks in the Army during the Korean War "documented the importance of the 'tipping point,' the level of integration beyond which the white man feels his majority status endangered." Project Clear, done by Johns Hopkins University, concluded that the tension that occurred beyond this point could be controlled in the Army, but presciently warned that "in a situation of free choice, the indications from Clear suggested that most whites would flee when they felt their dominance threatened, at

Concern was voiced as early as 1974, when two scholars predicted that "the tipping point will operate in a gradual fashion in the military rather than in the dramatic threshold fashion of residential communities," but they warned that "the end result, nevertheless, could well be a significant diminishment of white recruits for the ground force units involved."[50] By 1978 the growing racial imbalance in the all-volunteer force was being attributed to the tipping effect, or in the words of Congressman Robin Beard, "white flight."[51] Army Secretary Martin R. Hoffman acknowledged in 1977 that the Army had to virtually rebuild its corps of noncommissioned officers "who were hostile to the increased number of blacks."[52]

But attributing the change in the racial mix to the tipping effect is largely conjectural. Reductions in the levels of real military pay and termination of the Vietnam-era GI Bill could also have been contributing factors. Indeed, one of the two scholars who initially predicted that the tipping effect would apply had second thoughts:

The degree to which the changing racial composition of the Army reflects white reluctance to join an increasingly black organization is unknown, though it is surely a factor. Yet, I am unpersuaded that any significant number of middle-class whites—or middle-class any race, for that matter—would be more likely to join the Army, under present recruitment incentives, even if the Army was overwhelmingly white.[53]

But even if the tipping theory can be discounted, a related phenomenon with potential consequences for effectiveness has been described:

What may be happening in the all-volunteer Army, I suggest, is something like the following. Whereas the black soldier is fairly representative of the black community in terms of education and social background, white entrants of recent years are coming from the least-educated sectors of the white community. My stays with Army line units also leave the distinct impression that many of our young enlisted white soldiers are coming from non-metropolitan areas. I am even more impressed by what I do not often find in line units—urban and

---

which point a segregated pattern would recur." For a summary of Project Clear, see Leo Bogart, ed., *Social Research and the Desegregation of the U.S. Army: Two Original 1951 Field Reports* (Markham, 1969).

50. Morris Janowitz and Charles C. Moskos, Jr., "Racial Composition in the All-Volunteer Force," *Armed Forces and Society,* vol. 1 (November 1974), p. 113.

51. Testimony by Representative Robin L. Beard, in *Status of the All-Volunteer Force,* Hearing before the Subcommittee on Manpower and Personnel of the Senate Committee on Armed Services, 95 Cong. 2 sess. (GPO, 1978), p. 69.

52. *New York Times,* January 11, 1977.

53. Charles C. Moskos, Jr., "Symposium: Race in the United States Military," *Armed Forces and Society,* vol. 6 (Summer 1980), p. 593.

suburban white soldiers of middle class origins. In other words, the all-volunteer Army is attracting not only a disproportionate number of minorities, but also an unrepresentative segment of white youth, who, if anything, are even more uncharacteristic of the broader social mix than are our minority soldiers. Though put far too crassly, there is an insight in the assessment given me by a longtime German employee of the U.S. Army in Europe: "In the volunteer Army you are recruiting the best of the blacks and the worst of the whites."[54]

Available data support this view. As was shown in chapter 3, the average black Army recruit has a higher level of education than his or her white counterpart, and minorities, especially blacks, from "better" backgrounds and with "better" credentials are disproportionately attracted to the armed forces, as are minorities with higher educational aspirations. Added to this is the evidence that many of the youths entering the armed forces come from the most segregated areas of civilian life where the respective groups have had little previous contact with each other.[55] "It doesn't take long, for anyone visiting military bases and talking with the soldiers," writes James Fallows, "to see who they are and where they come from . . . white country boys, blacks and browns from the cities."[56] The trend has manifested itself in a number of ways, including a disturbing increase in Ku Klux Klan-type activities.[57] Though far from being widespread—by one estimate, military KKK members number about 200—the increase in klanism in the armed forces is viewed as a dangerous omen.[58]

54. Charles C. Moskos, Jr., "The Enlisted Ranks in the All-Volunteer Army," in John B. Keeley, ed., *The All-Volunteer Force and American Society* (University Press of Virginia, 1978), pp. 46–47.

55. John S. Butler and Kenneth L. Wilson, "The American Soldier Revisited: Race Relations and the Military," *Social Science Quarterly*, December 1978, p. 466.

56. James Fallows, *National Defense* (Random House, 1981), p. 127.

57. Racial incidents involving the KKK have been reported on the aircraft carrier *Independence* (Blaine Harden, "Sailors Wearing Sheets Create Racial Incident Aboard Aircraft Carrier," *Washington Post*, September 6, 1979), the supply ship *Concord* (John Stevenson, "Navy Ships' Racial Tension Is Under Guarded Control," *Norfolk Virginian-Pilot*, November 11, 1979), and the carrier *America* ("KKK Activity Investigated Aboard Atlantic Fleet Ships," *Washington Star*, July 1, 1979). The resurgence of Klan activities has not been confined to the Navy. Reports of activities at Fort Hood, Texas, and Fort Carson, Colorado, have also been carried in the press (see *Army Times*, August 13, 1979), and klanism among Army troops in Europe was reported in the *St. Louis Post-Dispatch*, December 7, 1980.

58. Lothar H. Wedekind, "GIs in the Klan: A Look Under Their Hoods," *Air Force Times Magazine*, July 7, 1980, p. 5. The increase has not been confined to the armed forces. A rash of incidents—"from name-callings to cross-burnings to physical attacks"— was being reported in the fall of 1980 on a number of college campuses across the country. Black students were reportedly anxious about "the resurgence of organized activity by

Studies of the formation of friendships among Army recruits during basic combat training emphasize the point. Research has shown that trainees select friends from their own race 1.3 to 1.5 times as often as would be expected on a random basis. More important, in contrast to the theory that intergroup contact fosters better intergroup relations—the so-called contact hypothesis—recent research yielded this result: not only do recruits "favor same-race peers as friends . . . [but] the additional anticipation that this inclination would diminish with time was disconfirmed."[59] These patterns, it is contended, have important implications for intergroup social relations:

> Under favorable conditions, these social subunits serve as a buffer between the individual's cultural heritage and disparate norms and requirements existing in the racially disparate unit. Under conditions unfavorable to good intergroup relations, however, the ties to these racially defined cliques within the unit may gain precedence over those to the unit itself, causing the unit to change from a cohesive group to a loose confederation of racially and culturally homogeneous subgroups.[60]

One possible "unfavorable" condition—committing a military unit to action in a situation that might test racial allegiances—has worried some observers.

Allegiances

That black soldiers would prove unreliable should they be called upon to take up arms against their "brothers" in either a domestic civil disorder or a foreign action—a charge that is understandably reprehensible to blacks—has long been the subject of speculation.

On the domestic side, concern was expressed during the 1960s: "that black soldiers may find they owe higher fealty to the black community than to the United States Army is a possibility that haunts commanders."[61] This fear was sparked by an incident involving the "Fort Hood 43," a group of black soldiers of the Army's 1st Armored Division at

---

hate groups, victories by right-leaning politicians, attempts in Congress to halt the enforcement of school busing and to repeal the Voting Rights Act, unsolved murders of blacks in such cities as Atlanta and Buffalo, and the acquittal of whites involved in the deaths of blacks in Miami and Greensboro, N.C." See Lorenzo Middleton, "New Outbreak of Cross-Burnings and Racial Slurs Worries Colleges," *Chronicle of Higher Education,* January 12, 1981, pp. 1, 12.

59. Francis E. O'Mara, "Affiliative Processes in Military Units: Racial and Cultural Influences," *Youth and Society,* vol. 10 (September 1978), p. 93.

60. Ibid., p. 94.

61. Moskos, *The American Enlisted Man,* p. 128.

Fort Hood, Texas, who refused to deploy for riot duty at the Democratic National Convention in Chicago in 1968.[62] It has been contended that one of the reasons for the National Guard's lack of success in recruiting blacks in the late 1960s was "the black fear that Guard duty will require them to confront black civilians in an adversary situation."[63] On the other hand, in numerous situations in the late 1960s and early 1970s racially mixed military units—both active and reserve—were deployed in riot areas without major incident. In fact, the relative effectiveness of active Army troops in the civil disorder in Detroit in July 1967 was attributed by the Kerner Commission in part to "the higher percentage of Negroes in the Active Army" than in the participating units of the Michigan National Guard,[64] and it recommended substantial increases in the participation of blacks in the Army and Air Force National Guard.

Yet doubt about the behavior of federal troops in domestic situations with racial overtones persists and should not be casually dismissed. This was indicated in a 1975 survey of enlisted personnel in selected Army units,[65] which found that one out of every three soldiers—black or white—preferred units composed entirely of soldiers of his own race. When faced with a hypothetical situation of white-black domestic conflict, the segregationists of each race were the most willing to serve against the opposite race and the least willing to be employed against their own race, as shown below in the percentages of soldiers who indicated they would volunteer or willingly follow orders to serve in the situation described:[66]

---

62. The group included twenty-six Vietnam veterans. According to press accounts, one of the veterans said: "We shouldn't have to go out there and do wrong to our own people. I can't see myself spraying tear gas on my fellow people." And an Army official was quoted as saying, "The problem is so fearful that we won't even discuss these people as Negroes." See *Time*, September 13, 1968.

63. Adam Yarmolinsky, *The Military Establishment: Impacts on American Society* (Harper and Row, 1971), p. 344.

64. *Report of the National Advisory Commission on Civil Disorders* (GPO, 1968), p. 276.

65. Charles C. Moskos, Jr., and Col. Charles W. Brown, U.S. Army, "Race Attitudes and Military Commitments in the All-Volunteer Army," paper prepared for the Biennial Conference of the Inter-University Seminar on Armed Forces and Society, October 16–18, 1975. Moskos and Brown surveyed 192 whites and 121 blacks from four battalions whose racial composition broke down as follows: infantry, 63 percent black; armored, 57 percent black; airborne, 40 percent black; ranger, 20 percent black (p. 2).

66. Ibid., p. 10.

	White		Black	
Situation	Integra-tionists	Segrega-tionists	Integra-tionists	Segrega-tionists
Stop violence of whites threatening integration of schools	80.2	56.4	78.9	82.9
Stop violence of blacks threatening private property	93.1	96.3	60.6	37.1

While the investigators stressed "that a clear majority of both white and black soldiers stated they would be willing to maintain the peace in situations of racial conflict," they also pointed out that even the integrationists (those who preferred racially mixed units) exhibited signs of the tension that could arise if they were called upon to oppose members of their own race in domestic conflicts.[67] The authors concluded that "domestic control of race riots is a mission which the regular Army must seek to avoid to the maximum extent possible."[68]

A related question is black soldiers' allegiance in some areas of international disorder. Generally, blacks' ambivalence about participation in a foreign conflict dates back at least to the late nineteenth century. According to one scholar:

From the outset of Cuba's struggle for liberation from Spain, black Americans expressed great sympathy with the rebel cause. Pronouncements supporting the insurrection emphasized the considerable black population in Cuba and the role played by black soldiers in the Cuban revolutionary armies.[69]

Then, as now, the issue was equality and citizenship. "In previous wars, even wars that were unpopular among Black Americans . . . the dominant attitude in Black communities was to use participation in the war to reenforce claims to equality and full citizenship."[70] Thus even while black troops fought patriotically, many other blacks failed to see the logic of fighting for abstract democratic principles in Cuba when they were denied these simple liberties at home. The extent of ambivalence

67. Ibid., p. 13. The possibility that white troops might balk at orders to quell disturbances in which white civilians are involved has generally been lost in the discussion of racial allegiance. A case in point was the use of federal troops in Little Rock, Arkansas, in 1957, when a unit of the 101st Airborne Division was deployed to enforce integration of public schools.

68. Ibid., p. 14.

69. Jack D. Foner, *Blacks and the Military in American History: A New Perspective* (Praeger, 1974), p. 72.

70. Robert W. Mullen, *Blacks in America's Wars: The Shift in Attitudes from the Revolutionary War to Vietnam* (Monad Press, 1973), p. 35.

among blacks on this issue and the concern that it generated were evidently sufficient to compel Booker T. Washington to assert that blacks "will be no less patriotic at this time than in former periods of storm and stress . . . [the black man] was an American through and through, and the President need not fret about divided allegiance, because there were no hyphenates among us."[71]

After the war in Cuba, blacks participated in several campaigns in the Philippines. Throughout, blacks in uniform expected that "their valor and loyalty would earn the appreciation of whites and lead to an easing of the oppression of blacks."[72] Their hopes were dashed, however, as the conclusion of war coincided with one of the most racially repressive eras in the nation's history. Cognizant of the increasingly militant posture of black organizations at home as well as the lack of attention paid to their performance on the battlefield, black soldiers "became less obsequious in manner, increasingly imbued with racial pride, less disposed to accept discriminatory treatment and physical violence, and resentful of ostensible praise that was couched in patronizing terms and racist innuendos."[73]

The black population in general was opposed to U.S. involvement in the Philippines. The black press and black intelligentsia strongly supported Filipino independence and opposed American colonization of nonwhite peoples. Moreover, it was felt that U.S. intervention in the Philippines was bound to have an adverse effect on black-white relations at home.

These circumstances put black troops in an uncomfortable position. It was reported that they identified with the plight of the nonwhite Filipinos, who, like themselves, were the targets of racial epithets. Black soldiers also believed, however, that superior performance on the battlefield would yield positive results at home. The feeling at home deepened their ambivalence: "Opposition to the war by Blacks became so loud that by 1899 the War Department questioned whether it would be wise to send any Black troops at all to the islands."[74] One official of the War Department wondered whether black troops "if brought face to face with their colored Filipino cousins could be made to fire on them."[75]

71. Foner, *Blacks and the Military*, p. 73.
72. Ibid.
73. Ibid., p. 74.
74. Mullen, *Blacks in America's Wars*, p. 39.
75. Willard B. Gatewood, Jr., *"Smoked Yankees" and the Struggle for Empire: Letters from Negro Soldiers, 1898–1902* (University of Illinois Press, 1971), p. 13.

Black troops did fight in the Philippines, but not without misgivings on their part and that of black civilians.

It was not until the United States became deeply involved in Southeast Asia in the 1960s that the issue again became prominent. The social divisiveness of that period was not principally a racial phenomenon. Yet the civil rights and black power movements became inextricably intertwined with the war largely because the conflict diverted attention and resources from civil rights initiatives and the War on Poverty.[76]

Although most leaders in the black community opposed the war, some were more strident than others. According to one source, with the exception of relative conservatives such as Roy Wilkins (NAACP) and Whitney M. Young (National Urban League), major black leaders subscribed to the view that

the stated war aims of the United States were hypocritical, that a bond of color existed between Black Americans and the yellow Vietnamese, that the United States was capable of committing genocide against a nonwhite people, and that war spending, when contrasted with spending to improve conditions of America's Black and white poor, illustrated the inhuman priorities of American society.[77]

Malcolm X was propounding the view that American involvement in Vietnam perpetrated violence against people of color while blacks at home were denied the rights they were seeking to obtain for others on foreign soil. In 1966 the Black Panthers proclaimed that blacks "should not be forced to fight in the military service to defend a racist government that does not protect us. We will not fight and kill other people of color in the world who, like Black people, are being victimized by the white racist government of America."[78] Of course, at the same time many white groups were opposing U.S. involvement in Vietnam, though on different grounds; indeed, the fealty of many white youths was at least as suspect as that of black youths.[79]

76. See Daniel P. Moynihan, *Maximum Feasible Misunderstanding: Community Action in the War on Poverty* (Free Press, 1969).

77. Mullen, *Blacks in America's Wars,* p. 64.

78. Ibid., p. 20.

79. For example, the 121,000 men who sought to avoid military service as conscientious objectors in 1970–71 were described by a former staff member of the Selective Service System as follows: "On the average, they had almost fifteen years of formal education (three years more than high school); more than four of ten were college graduates; more than seven of ten had some college education, and only one of twenty-five had not graduated from high school. Very few blacks or other minority representatives appeared among their ranks." Kenneth J. Coffey, *Strategic Implications of the All-Volunteer Force: The Conventional Defense of Central Europe* (University of North Carolina Press, 1979), p. 7.

Doubts about the allegiance of black soldiers nonetheless seemed legitimate at the time and to some whites remain so today. These doubts have been reinforced by sociological research. An analysis done in 1976 of the attitudes of Army paratroopers toward foreign combat missions, for example, indicated that, while most of the soldiers, regardless of race, showed a willingness to participate in combat, in certain situations nonblacks apparently would be more willing to participate than blacks: "rescuing American civilians who are in danger in an overseas country"; "an overseas war that the American people wholeheartedly support"; "protecting installations in an overseas country which are vital to America's economic needs—say, oil." In fact, race was the most significant variable in the first two scenarios. In two others, in which the troops seemed the most reluctant to engage—"a civil war in an overseas country in which the government asked for American help" and "an overseas war over which there is a lot of opposition at home"—race was not a significant factor. The investigators speculated that "what may possibly be implied is that blacks perceive the worth of American society in less positive terms and are correspondingly less willing to avow that they would volunteer in a situation where they might have to sacrifice their lives on behalf of that society."[80] But Margaret Bush Wilson, chairman of the board of the NAACP, characterizes the suggestion that "blacks may refuse to fight in certain parts of the world—or possibly not at all" as a "smokescreen thrown up by more subtle, sophisticated racists."[81]

In any event, the prospects for U.S. military involvement in regions of the world where racial allegiance might become a factor must be taken into account. The most obvious, though seemingly the most remote, would be a commitment of U.S. troops in support of the Pretoria government to suppress local strife in South Africa.[82] Less of a test would be posed by U.S. intervention, say, to thwart further Soviet meddling in the black nations of sub-Saharan Africa, inasmuch as any likely conflict would be between black factions. Oddly enough, Cuba

80. William C. Cockerham and Lawrence E. Cohen, "Volunteering for Foreign Combat Mission: An Attitudinal Study of U.S. Army Paratroopers," *Pacific Sociological Review*, vol. 24 (July 1981), pp. 329–54.
81. "All-Vol Critics Accused of Racism," *Army Times*, August 4, 1980.
82. Even this situation may be changing. For a discussion of the expansion of the role of blacks in the South African Defense Force, see Kenneth W. Grundy, "Black Soldiers in a White Military: Political Change in South Africa," *Journal of Strategic Studies*, vol. 4 (September 1981), pp. 296–305.

went out of its way to assign blacks to its contingents in Angola in the mid-1970s. By one estimate, about half the Cuban troops in Angola were black, twice their representation in the general population. "This disproportionate mobilization of blacks was meant to reduce the racial differences between the Cubans and their Angolan allies."[83] But the issue is probably moot; if U.S. military forces can play any useful role in the racial and other confrontations that can be foreseen in Africa, they will probably be limited to providing logistical support to United Nations peacekeeping forces brought in to separate the belligerents. And such UN contingents would almost certainly be made up of troops from the smaller powers.

In the politically volatile Middle East and Southwest Asia, a combination of important U.S. interests and the lack of reliable indigenous forces allied to the United States suggests a more credible prospect for military involvement, but any implications this might have for racial divisiveness in U.S. military forces are difficult to assess. The courtship of Arab leaders by several black American spokesmen during the Iran crisis was probably prompted as much by deteriorating black-Jewish relations in the United States as by racial solidarity between blacks and Arabs. Racial bonds between American blacks and Arab people are far from strong. The connection between black Muslims and Islam could be an important factor, particularly if the influence of the Muslims in the American black community should grow. Yet the Arab nations are not without their own racial hierarchies, and strain between the Arabs and black Africans is long-standing.

Latin America is also a region in which the racial connotations of military intervention cannot be completely discounted, especially in the Caribbean and Central America. That the United States may be willing to commit ground forces to contain communist expansion in these areas was indicated in the early days of the Reagan administration.[84] Moreover, any prospect of military action in Latin America is likely to precipitate

83. Jorge I. Domínguez, *Cuba: Order and Revolution* (Belknap Press, 1978), p. 354. Also, Rand researchers contend that many Soviet Central Asian troops were deployed in Afghanistan in 1979–80 "to blunt the political impact of the violent invasion." According to this account, the public relations effort was a failure; widespread fraternization between the Soviet Central Asians and the Afghan people led the Soviet Union to replace the Central Asians with Slavic troops beginning in February 1980. S. Enders Wimbush and Alex Alexiev, *Soviet Central Asian Soldiers in Afghanistan* (Rand Corp., 1981).

84. Developments in El Salvador in early 1981 prompted Secretary of State Alexander M. Haig, Jr., to speak of drawing the line in El Salvador against Soviet expansionism and Moscow's use of Cuban forces in Latin America. *New York Times*, February 7, 1981, Following this line, Edwin Meese III warned Cuba and the Soviet Union that the Reagan

more concern about the growing Hispanic contingent in the armed forces than about its black members. However, U.S. military action in the Dominican Republic in 1965 provides an important counterexample. Failed attempts by the rebel forces in the Dominican Republic to persuade black U.S. soldiers to "turn your guns on your white oppressors and join your Dominican brothers" attests to the speculative nature of the concern.[85]

The likelihood of racial divisiveness in the U.S. armed forces in future military ventures may thus depend more on the degree of public support for the particular war than on the racial characteristics of the adversary. A more important ramification would be a possible reluctance on the part of national leaders, fearful of the consequences, to commit military units with a high proportion of minorities to domestic or overseas contingencies. This would reduce both the flexibility of decisionmakers and the ability of the nation's military forces to meet national security commitments.[86]

Foreign Perceptions and Reactions

In many respects, the effectiveness of armed forces should be measured as much by their ability to deter the military adventures of an

---

administration "will take the necessary steps to keep the peace any place in the world, and that includes El Salvador." *New York Times*, February 23, 1981. However, the administration would probably encounter formidable problems in marshaling public support for such a venture. Early in 1982, as tension mounted in Central America, 89 percent of respondents to a *Newsweek* poll opposed the use of U.S. troops to aid the government of El Salvador. *Newsweek*, March 1, 1982, p. 19.

85. Moskos, *The American Enlisted Man*, p. 130.

86. Some allege that the British chose not to intervene militarily in Rhodesia in 1965 for just this reason. Fears that British soldiers would refuse to fight "kith and kin," so the account goes, closed off the military option. (Reports even circulated that Lord Mountbatten threatened to resign if the decision was made to invade Rhodesia.) But according to a close observer, "one suspects that the kith and kin factor was inflated to cover a general aversion to a military solution arising from other more compelling reasons." Robert C. Good, *U.D.I.: The International Politics of the Rhodesian Rebellion* (London: Faber and Faber, 1973), p. 61. It is no secret that the British resisted as long as possible the deployment of Irish regiments—such as the Irish Guards or the Royal Irish Rangers—to Northern Ireland. And an unofficial report states: "In the early 1960's the army's 82nd Airborne Division was ordered by 'high political authority' to remove black soldiers from certain units committed to civil rights-related riot control duty. In the view of officers in the division in 1975, the decision had been based on the erroneous judgment that the black troopers could have perceived a greater loyalty to the rioters than to the army." Coffey, *Strategic Implications of the All-Volunteer Force*, p. 76.

enemy as by their ability to defeat that enemy. Moreover, the importance of the perceptions of allies and nonaligned nations cannot be ignored. What signals are conveyed by manning U.S. armed forces with a growing proportion of minorities? What reactions are invited? Much would depend on the nation and its societal views.

Adversaries

If any nation has a greater problem with issues involving minority groups than the United States, it is its principal adversary, the Soviet Union. That nation encompasses between 100 and 140 different nationality and language groups. Currently, the Slavic and Baltic nationalities— the so-called European Russians—constitute about 80 percent of the Soviet population. These people generally speak Russian and are urbanized and technologically sophisticated. The rest of the population—the Central Asian and Caucasian nationalities—are more rural, not as well educated, and with a weak command of the Russian language, the only authorized medium of communication within the armed forces.[87] Under current practice, according to refugee accounts, most Central Asians are assigned to construction, supply, and rear-area service functions; only a few find their way into high-priority units. This situation has obviously caused stress and strain within the Soviet armed forces.[88]

Not unlike the situation in the United States, participation by non-European Russians in the Soviet armed forces will probably increase as the effects of current ethnodemographic trends are felt. By one estimate, the birthrate of the Central Asians (defined as the four republics of Uzbekistan, Turkmenistan, Kirgizia, and Tadzhikistan, plus Kazakhstan) is two and a half to three times the national average.[89] At this rate, by the end of the century non-Europeans (mainly Central Asians) will constitute 20 to 25 percent of the Soviet population and close to 40 percent of its young adults.[90] One can only speculate about how the USSR will cope with what some Soviet officials privately refer to as the

   87. It is estimated that only 16 percent of Central Asians speak Russian with some fluency. See Jeremy Azrael, *Emergent Nationality Problems in the USSR*, R-2172-AF (Rand Corp., 1977), p. 18.
   88. Including reports on "the difficulty of preparing training manuals in different national languages." Ibid., p. 21.
   89. Rex D. Minckler, Robert N. Ginsburgh, and Richard G. Rebh, *Soviet Defense Manpower*, GE77TMP-18A (Washington, D.C.: General Electric Center for Advanced Studies, 1977), p. E-6.
   90. Azrael, *Emergent Nationality Problems*, p. 5.

"yellowing" (*ozheltenie*) of the Red Army. Although it would make economic sense to fill the ranks of the military with Central Asians, according to one scholar, this option will probably be rejected since it would mean assigning Central Asians to high-priority military units, thus extending to those units the command, control, and communications problems that thus far have been confined to the low-priority organizations. These problems could be significantly alleviated by a return to national military formations—Uzbek, Tartar, Kazakh, and so on—but the top leadership is unlikely to endorse any decisive move in this direction because of its "fear that indigenous units might provide tacit or open military support for nationalist challenges to central authority."[91]

Thus the Soviet Union better than any other major power understands and appreciates the potential problems associated with military integration issues. To the extent that it perceives that a greater reliance on "non-Russians" weakens its military machine—and there are signs that it does—it is just as likely to perceive that a greater reliance on racial minorities by the U.S. military somehow diminishes the fighting prowess of this nation's armed forces.

Potential adversaries might also view the changing racial balance as an opportunity to exploit racial problems where they exist and to create them where they do not. There is a long record of the nation's perceived vulnerability to such propaganda measures; virtually every recent adversary has used them. At the turn of the century, Filipino guerrillas exhorted U.S. black soldiers to desert and not be "instruments of their white masters' ambitions to oppress another 'people of color.'"[92] In World War I the Germans circulated among members of the black 92nd Division leaflets pointing out the contradiction of fighting for democracy abroad while being denied rights at home.[93] In World War II the Japanese made radio appeals specifically to black troops serving in the Pacific theater. During the Korean War the Chinese reportedly used "divide and conquer" techniques on black prisoners of war. In Vietnam the

91. Ibid., pp. 19, 22. Some American observers, apparently placing great stock in ethnic divisiveness in the Soviet Union, have suggested that the United States concentrate its nuclear targeting on regions where "the population is predominantly ethnic Russian in order to limit the damage in the non-Russian republics." The prospect that minority elements might inherit whatever remained of the Soviet Union in the wake of a nuclear riposte by the United States, according to this argument, might help deter the launching of a Soviet first strike. See Maxwell D. Taylor, "A New Measure for Defense," *Washington Post*, January 14, 1982.
92. Mullen, *Blacks in America's Wars*, p. 40.
93. Ibid., p. 48.

National Liberation Front announced that "liberation forces have a special attitude toward American soldiers who happen to be Negroes."[94] Rebel forces in the Dominican Republic, as already indicated, appealed to racial differences. And more recently, the Khomeini government released thirteen U.S. hostages after three weeks of captivity, eight of whom were black males and five white women. The release was staged as a major media event at which Khomeini's professed respect for women and oppressed blacks was highlighted.[95]

Allies

Ironically, more of the nation's allies than its potential adversaries may view the changing racial balance in the U.S. armed forces as a sign of weakness. Although their leaders are usually careful in their choice of terms, the attitudes of the West Germans are widely recognized. Strain between the German people and U.S. service personnel in general and black GIs in particular, which has long been evident, took a turn for the worse in the 1970s with the apparent deterioration in the "quality" of American troops. In response to pressure by the United States to share more of the military burden for the defense of Western Europe, a high-level German official, in a thinly veiled criticism of U.S. military personnel, said that "West German soldiers did not use drugs and could read and write."[96] A former high government official of another ally, Israel, was more direct, though inaccurate: "The United States has a very difficult problem concerning quality . . . up to the rank of sergeants. Most of the soldiers are black, who have a lower education and intelligence."[97]

Such negative perceptions among America's allies are significant not only because they tend to diminish the mutual confidence so necessary to successful military alliance relationships, but also because they tend to affect the morale and hence the performance of minority troops adversely. Although there are probably many factors that explain the substantially higher proportion of blacks assigned to units in Korea than in Germany (see table 4-6), differences in the two nations' attitudes toward black Americans must be a major influence.

94. Quoted in Moskos, *The American Enlisted Man*, p. 130.
95. "The Mounting War of Nerves," *Newsweek*, December 3, 1979, pp. 46–50.
96. Comment attributed to Finance Minister Hans Mathofer, *New York Times*, November 8, 1980.
97. Comments attributed to Moshe Dayan on Israeli television. Carl T. Rowan, "Dayan's Racist Remarks," *Washington Star*, December 5, 1980.

Emerging Nations

Relations with emerging nations, particularly those in mineral-rich sub-Saharan Africa, are of no small importance in U.S. foreign policy. Whether the image these nations have of the racial situation in the U.S. armed forces would be a plus, a minus, or even play any part at all in shaping their attitudes toward the United States is highly conjectural. Much would depend on the particular nation, its colonial history, its contemporary society, and its self-perception. It is risky to generalize because of the vast differences among the many nations involved; attitudes could run the gamut of possibilities.

On the one hand, the self-confident and relatively independent government of Nigeria, which has emerged as the champion of southern African liberation, would be inclined to view the racial composition of the U.S. military from a civil rights rather than a strategic military perspective. To that extent, like many black Americans, it would probably look with favor upon the expanded role of blacks in the armed forces but with concern upon the slower progress being made in integrating the officer corps. The francophone bloc of African nations, on the other hand, which is less inclined to identify with civil rights causes, might assess the situation from a military standpoint. Reflecting their past dependence on white military leadership, those nations might view a blacker American military as a weaker American military. In any case, too much should not be read into the speculation. There is little evidence that African nations are aware of, much less sensitive to, the shift in the racial composition of the U.S. ground forces.

THE EFFECTIVENESS of military forces depends on the capabilities of the individual members, the performance of the group to which they belong, and the image they present to the nation's allies and adversaries and to its citizens. A healthy measure of uncertainty remains about how the racial composition of the armed services affects all three. Much of the uncertainty derives from the problems inherent in measuring military capabilities in general and in ascertaining the perceptions of Americans, as well as those of the citizens and officials of foreign nations. While the evidence presented above is too thin to lead to conclusions about many aspects of the question, the discussion can identify gaps in our knowledge and understanding of the issues. Until those gaps are closed, predicting the implications, positive or negative, of the racial mix for American military effectiveness will remain highly speculative.

CHAPTER SIX

LOOKING AHEAD

THE ISSUES taken up in the previous chapters deserve a position high on the nation's public policy agenda. A look into the future—at trends that will make it increasingly difficult to man the military, on the one hand, and at the manpower policy options available to counteract those trends, on the other—reinforces the importance of the issues.

The racial composition of the armed forces in the 1980s and beyond will be shaped by many forces, of which some are uncontrollable, others unpredictable, and most not well understood. Changes in the composition of the population and in the state of the economy will affect the racial mix, as will planned increases in the size of the nation's military forces. Also bearing on the question are technological trends that will make weapon systems more complex and indications that the ability of American youth to operate and maintain those systems may be on the decline. Finally, changes in qualitative standards, military pay and benefits, and recruitment methods could have an important influence on the racial mix.

Ethnodemographic Trends

Dwindling birthrates in the United States—a trend that started in the late 1950s and brought the baby boom to an end in the mid-1960s—are having a significant impact on many areas of public policy. As the children of that period have grown older, the effects have already been felt, most notably by the nation's primary and secondary educational institutions. As the first cohorts of that generation complete high school, starting about 1983, higher educational institutions and the civilian labor force will begin to notice the effects, as will the armed forces, which have traditionally attracted eighteen- to twenty-one-year-old volunteers.

Table 6-1. Projected U.S. Population Aged 18 to 21, by Sex and Race, Selected Years, 1981–95
Thousands

Category	1981	1983	1985	1987	1989	1991	1993	1995
Male	8,618	8,356	7,821	7,356	7,404	7,197	6,702	6,608
White	7,281	7,010	6,509	6,085	6,098	5,864	5,405	5,331
Black	1,147	1,145	1,102	1,053	1,070	1,071	1,022	994
Other	190	201	210	218	236	262	275	283
Female	8,401	8,142	7,621	7,164	7,197	6,984	6,495	6,386
White	7,059	6,799	6,312	5,896	5,897	5,666	5,220	5,137
Black	1,168	1,161	1,116	1,067	1,081	1,076	1,022	990
Other	174	182	193	201	219	242	253	259
Total	17,019	16,498	15,442	14,520	14,601	14,181	13,197	12,994

Source: Bureau of the Census, *Current Population Reports,* series P-25, no. 704, "Projections of the Population of the United States: 1977 to 2050" (Government Printing Office, 1977), pp. 40–60. Figures are rounded.

The magnitude of the change can be seen in table 6-1; compared to 1981 levels, there will be about 15 percent fewer in this age group by 1987 and 24 percent fewer by 1995.[1]

Also relevant is the shift that will occur in the racial characteristics of the young population. In contrast to 1981, when black males constituted 13.3 percent of the total males in the age group, by 1989 the percentage will be 14.5 and by 1995 it will have increased to 15 percent. During the same period, the group of males categorized as "other minorities" will increase from 2.2 percent of the total in 1981 to 4.3 percent in 1995.

Tested Abilities and Aptitudes

As discussed in chapter 5, by present measures minority youths tend to have less aptitude than white youths for general military duty and even less for technical and mechanical skills. The ethnic shift in the youth population, then, would be expected to reduce the overall proportion of those of military age who would ordinarily qualify for enlistment and training in high-skill jobs. The extent of this effect, however, would be influenced by changes in the relative abilities of black and white youths.

1. Cohort size will begin to increase again after 1995, since the annual number of births began to rise in 1975. This is not because fertility rates have risen but rather because of the delayed "echo effect": there are simply more women, born during the baby boom, who are now of childbearing age. Once this generation passes that age, starting in the 1990s, annual births will probably decline once again. See Bryant Robey, "Fertility Fantasies and a Few Facts," *American Demographics,* January 1981, p. 2.

In recent years the abilities of the younger generation of Americans, especially their intellectual capacity and basic academic skills, have caused concern. The most widely cited evidence of a general downward trend is the monotonic decline in Scholastic Aptitude Test (SAT) scores since 1963. In that year the average score on the verbal section of the SAT was 478 and on the mathematics portion was 502; in 1979 the verbal mean score was down to 427 and the mathematics score to 467—the lowest in the history of the test.[2] The possible causes of this downward trend in achievement have been widely discussed and are documented in the literature.[3] Whether the decline has been more prominent in one population subgroup or another is not discernible since the racial and ethnic status of test participants is not available.

Racial comparisons can be drawn, however, from scores attained by college-bound high school students on tests administered by the American College Testing program. Test score data, arrayed in table 6-2, show a general downward trend during the 1970s, with most of the decline occurring during the first half of the decade. These data, however, failed to reveal any significant difference in relative abilities by race; the decline in the composite test scores of blacks generally ran parallel to the decline of the entire student cohort. The patterns of change were similar for the subtests with the exception of the English subtest, on which blacks registered a relative gain.

In what could be a related development, relative gains by black youths are also evident in reading ability assessments by the National Assessment of Educational Progress. Measurements taken from 1971 to 1980 show a slight but statistically insignificant decline in the mean total scores obtained by seventeen-year-olds on tests of literal comprehension, inferential comprehension, and reference skills, with whites registering a small decline and blacks a slight increase (see table 6-3). While

---

2. "SATs Drop Again," *Report on Educational Research*, September 19, 1979, p. 3.

3. See, for example, Willard Wirtz, ed., *On Further Examination: Report of the Advisory Panel on the Scholastic Aptitude Test Score Decline* (New York: College Entrance Examination Board, 1977); Yvonne L. Wharton, *List of Hypotheses Advanced to Explain the SAT Score Decline* (Princeton, N.J.: College Entrance Examination Board, 1977); Hunter M. Breland, *The SAT Score Decline: A Summary of Related Research* (Princeton: CEEB, 1976); Bernard Rimland and Gerald E. Larson, "The Manpower Quality Decline: An Ecological Perspective," *Armed Forces and Society*, vol. 8 (Fall 1981); and Brian K. Waters, "The Test Score Decline: A Review and Annotated Bibliography," Technical Memorandum 81-2 (Office of the Secretary of Defense, August 1981).

Table 6-2. Comparative Means and Standard Deviations of American College Testing Program Assessment Scores of College-bound Students, Selected School Years, 1970–80

Test category	1970–71		1974–75		1979–80	
	Total	Black	Total	Black	Total	Black
English						
Mean	18.0	12.2	17.7	13.0	17.9	12.7
Standard deviation	5.5	5.4	5.3	5.1	5.4	5.0
Mathematics						
Mean	19.1	12.9	17.6	10.5	17.4	10.8
Standard deviation	7.1	5.5	7.9	6.4	7.6	6.1
Social studies						
Mean	18.7	11.8	17.4	10.6	17.2	10.8
Standard deviation	7.0	6.3	7.5	6.3	7.3	5.9
Natural science						
Mean	20.5	14.5	21.1	14.9	21.1	15.0
Standard deviation	6.3	5.0	6.3	4.9	6.2	5.1
Composite						
Mean	19.2	13.0	18.6	12.4	18.5	12.5
Standard deviation	5.5	4.4	5.8	4.7	5.8	4.6

Source: Data provided by the American College Testing Program, Iowa City, Iowa, July 1981.

the performance of this cohort remained at virtually the same level in literal comprehension (locating or remembering the exact meaning of a word, sentence, or paragraph) and reference skills (applying reading behavior to problem-solving), a significant decline, strongest among whites, occurred in inferential comprehension (gleaning from passages ideas that are not explicitly stated).[4] The study also found substantial improvement in the reading performance of nine- and thirteen-year-olds, especially among black children. While this could portend further gains by blacks as these cohorts age, it appears that the rate of achievement among students diminishes as both racial groups progress through their years of schooling.

The roughness of these data, particularly the racial comparisons, provide a shaky foundation on which to base projections. The leveling off in the decline of SAT scores in 1981 is a hopeful sign, as are gains in reading comprehension, but whether they mark the end of the downward

4. For a complete description of the test and results, see National Assessment of Educational Progress, *Three National Assessments of Reading: Changes in Performance, 1970–80,* Report 11-R-01 (Denver: Education Commission of the States, 1981).

Table 6-3. Mean Percentages and Changes in Correct Responses for Seventeen-Year-Olds in Reading Performance, by Race, 1971 and 1980[a]

Reading performance	1971			1980			Change in mean percent correct, 1971–80		
	White	Black	Total	White	Black	Total	White	Black	Total
All exercises	71.2	51.7	68.9	70.6	52.2	68.2	−0.7	0.5	−0.7
Literal comprehension	74.2	57.0	72.2	74.1	57.5	72.0	−0.1	0.5	−0.2
Inferential comprehension	66.5	47.0	64.2	64.5	45.9	62.1	−2.0	−1.1	−2.1
Reference skills	72.5	45.6	69.4	73.2	49.8	70.2	0.7	4.2	0.8

Source: National Assessment of Educational Progress, *Three National Assessments of Reading: Changes in Performance, 1970–80*, Report 11-R-01 (Denver: Education Commission of the States, 1981), pp. 57–60. Figures are rounded.
a. Includes only those in school.

trend remains to be seen.[5] At any rate, there is little else in recent test results to suggest that changes in scholastic or military aptitude, in either direction, will be more than marginal in the 1980s. As for racial comparisons, it seems reasonable to expect that the aptitude gap will narrow as blacks continue to realize the benefits of two decades of social and educational reform. But again, progress is likely to be slow since these programs cannot be expected to change so quickly what decades of neglect have perpetuated.

Economic Factors

The extent to which shrinkage in the pool of prospective volunteers causes recruitment problems will depend on the proportion of youths who choose military service over available civilian employment. In recent times, notably since the end of conscription, the military services have had more difficulty attracting recruits from the upper levels of the quality spectrum during periods of economic growth and declining rates of unemployment. Alternatively, when the labor market has slackened, the number and quality of applicants have increased. How much the changes in recruitment can be attributed to changes in the labor market is a contentious issue; estimates of unemployment elasticities range from 0.2 to 0.5. Thus a change of, say, 10 percent in the youth unemployment rate would give rise to a 2 to 5 percent change in enlistments of high-quality volunteers—defined as male high school graduates scoring above the thirtieth percentile on the standardized entry test.[6]

This aggregate estimate, however, masks important differences by

---

5. The Department of Education is optimistic that the fifteen-year decline in educational achievement has bottomed out, particularly among traditionally low-achievement groups. In addition to improvements in the educational status of minority-group nine-year-olds as measured by the National Assessment of Educational Programs, the department cites various gains in educational achievement in New York City, Chicago, and New Jersey. Department of Education, "The Annual Evaluation Report," Secretary's Summary, December 2, 1981.

6. Congressional Budget Office, *The Costs of Defense Manpower: Issues for 1977*, Budget Issue Paper (Government Printing Office, 1977), p. 135; David W. Grissmer, "The Supply of Enlisted Volunteers in the Post-Draft Environment: An Analysis Based on Monthly Data, 1970–1975," in Richard V. L. Cooper, ed., *Defense Manpower Policy: Presentations from the 1976 Rand Conference on Defense Manpower*, R-2396-ARPA (Rand Corp., 1978), p. 110; CBO, *Costs of Manning the Active-Duty Military*, Staff Working Paper (GPO, 1980), p. 38; and Lawrence Goldberg, "Navy Enlisted Supply Study," CNA 81-0601 (Alexandria, Va.: Center for Naval Analyses, April 9, 1981).

race. The studies that have attempted to distinguish between the responses of white and minority youth to changes in unemployment rates have consistently measured a significantly smaller response from blacks; in some cases, the change has been in the *opposite* direction. While a 10 percent increase in white youth unemployment, by one estimate, would result in roughly a 5 percent increase in high-quality white volunteers, a similar increase in black youth unemployment would yield a 6 percent *decrease* in black volunteers with similar qualitative characteristics.[7] This anomaly is probably a manifestation of the "substitution effect," whereby during periods of rising unemployment, for example, the services have been able to attract more high school graduates that scored in the top two AFQT categories and therefore have been less inclined to accept volunteers from the bottom two categories. The fact that white youths are more likely to be in the former group and blacks in the latter supports this explanation.

The Economics of the 1980s

The economic climate of the 1980s will be a major factor in civilian employment opportunities for America's youth. A gloomy job picture will influence more young men and women to seek military service instead of entering the civilian labor market. Economic projections are always risky, but they have been made more uncertain by the Reagan administration's unorthodox economic program, which calls for large cuts in nondefense expenditures while substantially increasing defense outlays.[8] This shift in spending is to be accompanied by decreases in personal income tax rates and by tax incentives for businesses (for example, liberalizing depreciation allowances to spur capital formation). The administration "expects that these policies, together with a steady decline in the growth of money and substantial deregulation of the

---

7. Grissmer, "Supply of Enlisted Volunteers," p. 111. Estimates of the relation between enlistments and black youth unemployment are particularly shaky, given the quality of available data on black youths' earnings and employment. Most researchers have expressed misgivings about the size of the sample data and the statistical reliability of census surveys of minority youth.

8. For a general discussion of the Reagan administration's economic program as well as a detailed description of the proposed program cuts and their implications, see Joseph A. Pechman, ed., *Setting National Priorities: The 1982 Budget* (Brookings Institution, 1981).

Table 6-4. Comparison of Rates of Unemployment Projected by the Reagan Administration and the Congressional Budget Office, 1982–87
Percent

Projection by	Calendar year					
	1982	1983	1984	1985	1986	1987
Administration	8.9	7.9	7.1	6.4	5.8	5.3
Congressional Budget Office	8.9	8.0	7.4	7.2	6.9	6.7

Sources: *The Budget of the United States Government, Fiscal Year 1983* (GPO, 1982), pp. 2-5, 2-7; and Congressional Budget Office, *Baseline Budget Projections for Fiscal Years 1983–1987* (GPO, 1982), p. 6.

economy, will have a large, favorable effect on inflation and economic growth."[9]

The administration also expects its economic plan to lower the nation's unemployment rate from 8.9 percent in 1982 to 5.3 percent in 1987 (see table 6-4). The Congressional Budget Office, on the other hand, using somewhat different assumptions, estimates that the unemployment rate will fall by only 2.2 percentage points (from 8.9 to 6.7) during this period.

In the past a decline, say, of one percentage point in the overall unemployment rate has been accompanied by a decrease of 1.6 percentage points in the rate of youth unemployment. If that relationship should continue, the predicted 3.6 percentage point reduction in the nation's unemployment rate from 1981 to 1986 would lower the youth unemployment rate by 5.8 percentage points. Yet according to the CBO's projections, the unemployment rate of young people would fall by only 3.5 percentage points. In either case, the predicted decline in the youth unemployment rate, though welcome, would still leave it very high. Moreover, if past differences in youth unemployment along racial lines persist, blacks can expect even less relief than their white counterparts.

Factors Affecting Black Youth Unemployment

It is difficult enough to predict with accuracy how changes in the general economic climate in the 1980s will affect the employment opportunities of young black males and their consideration of military service as an alternative. But also clouding the picture are the uncertain effects of the relative occupational immobility of black youth, of changes

9. Congressional Budget Office, *Economic Policy and the Outlook for the Economy* (GPO, 1981), p. xxii.

in the structure of the labor force, and of possible changes in youth-oriented government programs.

The future decline in the number of young Americans can be expected to create more intense competition among institutions of higher education, vocational schools, private employers, and the military. Although this seems to indicate more favorable youth employment prospects, for young blacks in particular, who will constitute a larger proportion of the group, such optimism may be misplaced. Job opportunities will depend on the strength of labor market "segmentation" and on the extent of labor market "substitution."

SEGMENTATION. Black youths, like their white counterparts, are concentrated in unskilled, low-paying jobs in the service, manufacturing, and trade industries, but some contend that blacks are not able to move as easily as whites into other, nontraditional youth jobs.[10] Thus youths are not only segregated into a limited number of occupations, but also segmented racially. "If this interpretation is valid, such occupational segregation would hurt a large cohort of black youths entering the labor force unless the available occupations grew fast enough to accommodate all new workers."[11] Regardless of whether labor market segmentation has a racial element—and some contend that it does not[12]—the interaction between occupational segmentation and the number of black youths seeking jobs may still have an adverse effect on young blacks since there is a declining trend in the relative number of jobs for the young. In other words, the situation could become one of too many workers seeking too few jobs.

SUBSTITUTION. Also bearing on the question is the changing composition of the labor force, which could put youth at an even greater disadvantage than in the past. While the working population as a whole is growing older and becoming more experienced, the participation rate of women and immigrant workers (both legal and illegal) is increasing.

10. See, for example, Paul Osterman, "Racial Differentials in Male Youth Unemployment," in Department of Labor, Office of the Assistant Secretary for Policy, Evaluation, and Research, *Conference Report on Youth Unemployment: Its Measurement and Meaning* (GPO, 1978).

11. Robert D. Mare and Christopher Winship, "Changes in Race Differentials in Youth Labor Force Status: A Review of the Literature," in *Expanding Employment Opportunities for Disadvantaged Youth: Sponsored Research* (Washington, D.C.: National Commission for Employment Policy, 1979), p. 8.

12. One is Diane M. Westcott, "Youth in the Labor Force: An Area Study," *Monthly Labor Review* (July 1976), pp. 3–9.

This increasing competition could well reduce employment opportunities for young job seekers, especially blacks. Although there are research findings on the substitutability of women for youth in the labor market,[13] little empirical evidence exists on the extent to which immigrants (especially illegal ones) displace young workers in general and blacks in particular. Nevertheless, some contend that the "recent influx of adult immigrants willing to hold low-paying jobs" is a contributing factor to the perpetuation of the poor labor market status of black youths.[14]

National policy governing immigrant workers can therefore affect the rate of substitution. Under a proposal made by the Reagan administration, for example, a substantial number of Mexicans would be allowed to work in the United States on a temporary basis. This guest-worker program is apparently seen as a way both to improve U.S.-Mexican relations and to stem the tide of illegal immigration.[15] The impact of such a program on the prospects of young black males could be enormous. By most accounts, the overwhelming majority of illegal Mexican workers in the United States work in unskilled, low-wage, entry-level jobs—the same type as those held by many young blacks.[16] If the proposal increased the number of Mexicans in the United States and if those who would participate in a guest-worker program sought the same kinds of jobs as illegals, an added hardship would be imposed on job-seeking black youths.

YOUTH-ORIENTED POLICIES. A third factor that could affect the labor market status of black youth is the new course charted by the Reagan administration for many of the nation's social programs. In recent years, youth employment and training programs have grown in both number and scope. A major step, for example, was taken in 1977 with the passage of the Youth Education and Demonstration Projects Act (YEDPA), which established four new programs aimed primarily at disadvantaged

13. See, for example, Daniel Hamermesh and James Grant, "Do Employers Substitute Workers of Different Ages, Races and Sexes, and What Does This Imply for Labor Market Policy?" paper prepared for the National Commission for Employment Policy, October 1979.
14. Congressional Budget Office, *Youth Employment and Education: Possible Federal Approaches* (GPO, 1980), p. 6.
15. Robert Pear, "Reagan Aides Draft a Plan to Let Mexicans Work in U.S. as Guests," *New York Times,* May 11, 1981.
16. David S. North and Marian F. Houstoun, *The Characteristics and Role of Illegal Aliens in the United States Labor Market: An Exploratory Study* (Department of Labor, 1976).

youth and authorized discretionary funding for research on alternative ways of increasing the employability of all the young.[17]

Although many of the youth programs have experienced some difficulties with administrative and management procedures, most observers acknowledge that the programs have been instrumental in keeping the youth unemployment rate lower than it would have been without them. It is ironic, however, that one positive feature of the youth programs that could have deleterious future consequences is the heavy reliance of black youths on the programs for their primary employment. For example, between 30 and 40 percent of all black teenagers employed in 1978 were youth program participants.[18] This dependence leaves blacks highly vulnerable to the termination or cutbacks of the programs.

As part of a movement to streamline federal spending in 1981, the Reagan administration recommended that two youth programs (the Youth Conservation Corps and the Young Adult Conservation Corps) be terminated and that the Youth Employment and Training Program, the Youth Community Conservation and Improvement Project, and the Summer Youth Employment Program be combined with adult programs (at lower funding levels) under one employment and training grant to the states.

To offset the anticipated adverse effects of these reductions, consideration is being given to proposals for a subminimum wage for youth and for wage subsidies, both designed to encourage private employers to create more jobs for young workers. Proponents argue that a lower minimum wage for young workers, by reducing firms' cost of hiring them, will encourage an increase in their employment. But to effectively and efficiently administer a two-tier system would probably require more government regulation—a prospect that seems to run counter to the administration's general concern about regulation. Also, the fear has been voiced, largely by civil rights groups, that a subminimum wage would result in many young people taking jobs away from their parents.

Even if the many problems that would arise with a two-tier minimum wage system were solved, the employment relief that it would provide for youth—particularly blacks—would be only marginal, since other

---

17. These programs included the Youth Employment and Training Program, the Youth Incentive Entitlement Pilot Project, the Youth Community Conservation and Improvement Project, and the Young Adult Conservation Corps.

18. Testimony of George E. Johnson before the Joint Economic Committee, February 9, 1979.

factors (such as the general level of business conditions, demographic forces, and the degree of competition from other groups in the labor force) tend to influence youth employment more than the minimum wage.[19] Furthermore, "the minimum wage requirement is but one of several factors, such as payroll taxes, fringe benefits, and insurance costs, which have tended to raise the cost of employing workers, especially those with few skills and higher turnover propensity." Thus, "together, these additional costs may outweigh the minimum wage requirement in their influence on a business firm's decision to avoid hiring young people."[20] If overall improvement is only marginal, it stands to reason that the segment of the youth population with relatively more severe employment handicaps, which includes most black males, could not expect much, if any, relief.

Using wage subsidies to induce employers to hire more young workers is another strategy under consideration. As with the minimum wage, little is known about the effects of wage subsidies on youth employment.[21] For our discussion, the most relevant U.S. experience with wage subsidies for youth is under the existing Targeted Jobs Tax Credit (TJTC) program. Established in 1978, this program provides wage subsidies to employers who hire disadvantaged youth or targeted members of other economically disadvantaged groups.[22] Information on the use of wage subsidies is also available from the work incentive (WIN) tax credit program, originally passed in 1971 and expanded in 1975. The WIN program offers subsidies to firms that hire recipients of aid to families with dependent children who are registered in the WIN program. Unfortunately, the track record for both the TJTC and WIN programs is not good. In recent years neither has been well supported; many of the employers with eligible workers chose not to participate. According to one researcher:

Under the work incentive tax credit, credits claimed amounted to about 15,000 person years of employment; data from aid to families with dependent children surveys show actual employment of potentially eligible workers in the range of

19. See *Report of Minimum Wage Study Commission*, vol. 1: *Commission Findings and Recommendations* (GPO, 1981), p. 48.
20. *Employment and Training Report of the President: 1978* (GPO, 1978), p. 75.
21. See John L. Palmer, ed., *Creating Jobs: Public Employment and Wage Subsidies* (Brookings Institution, 1978).
22. In addition to poor youths, others eligible under the TJTC program include economically disadvantaged Vietnam-era veterans, supplemental security income recipients, general assistance recipients, economically disadvantaged ex-convicts, work-study youths, and vocational rehabilitation referrals.

500,000. Thus only a small fraction of employers claimed credits for which they were eligible. The cumulative number of newly hired disadvantaged youth for whom employers claimed the targeted jobs tax credit through March 30, 1980, was about 60,000. This figure is less than 4 percent of the nearly 2 million disadvantaged youth hired by private firms during the year.[23]

Several reasons have been offered to account for the failure of employers to respond to these programs. Those most often mentioned include (1) lack of knowledge about the size of the subsidy necessary to induce employers to hire targeted workers; (2) employers' aversion to the necessary paperwork to claim tax credits; and (3) the possible stigma associated with workers eligible for the program. Although the private sector's poor support of wage subsidies is still being debated, there is little empirical basis for optimism that such subsidies would improve the employment prospects of black youth to any significant extent.

In sum, despite the shrinking pool of American youth and the optimistic outlook for economic recovery, civilian employment opportunities for young black males are likely to deteriorate in the 1980s, especially if efforts to encourage private employers to create more jobs fail to offset expected setbacks in government job programs. Indeed, unless economic recovery is unusually vigorous, relatively more—not fewer—black youths can be expected to find military service an attractive option.

Technological Trends

Advances in technology since World War II have had a dramatic influence on the U.S. defense establishment.[24] Unlike the armed forces of an earlier period, which were dominated by relatively unskilled infantrymen and able-bodied seamen, the majority of military personnel today are involved in providing support for the combat forces.

The shift from work requiring general military skills toward tasks requiring special expertise is shown in table 6-5. The sharpest changes have occurred in the technical skills (those involving computer specialists, electronics technicians, medical technicians, and the like), the

23. Robert I. Lerman, "A Comparison of Employer and Worker Wage Subsidies," in Robert H. Haveman and John L. Palmer, eds., *Jobs for Disadvantaged Workers: The Economics of Employment Subsidies* (Brookings Institution, 1982), p. 177.

24. For a more complete discussion of the influence of technology on the military occupational structure, see Martin Binkin and Irene Kyriakopoulos, *Youth or Experience? Manning the Modern Military* (Brookings Institution, 1979).

Table 6-5. Distribution of Trained Military Enlisted Personnel by Major Occupational Category, All Services, 1945, 1957, and 1981
Percent

Major occupational category[a]	1945	1957	1981
White collar	28	40	46
Technical workers[b]	13	21	28
Clerical workers[c]	15	19	18
Blue collar	72[d]	60	55
Craftsmen[e]	29	32	28
Service and supply workers	17	13	11
Infantry, gun crews, and seamanship specialists	24	15	16

Sources: Data for 1945 and 1957 from Harold Wool, *The Military Specialist: Skilled Manpower for the Armed Forces* (Johns Hopkins Press, 1968), table III-3, p. 42. Data for 1981 provided by Defense Manpower Data Center. Figures are rounded.

a. Categories are based on the Defense Department occupational classification system.

b. Percentages for 1945 and 1957 include "electronics" and "other technical" categories. Percentage for 1981 includes "electronic equipment repairmen," "communications and intelligence specialists," "medical and dental specialists," and "other technical and allied specialists" categories.

c. Percentages for 1945 and 1957 include administrative and clerical personnel. Percentage for 1981 is for the "functional support and administration" category.

d. Includes 2 percent classified as miscellaneous.

e. Percentages for 1945 and 1957 include "mechanics and repairmen" and "craftsmen" categories. Percentage for 1981 includes "electrical/mechanical equipment repairmen" and "craftsmen" categories.

service and supply occupations, and the general combat skills. The proportion of technical jobs, which has always been higher in the equipment-intensive Navy and Air Force, markedly increased after World War II in the Army and, to a lesser extent, in the Marine Corps. For the Army, the increase in technical specialization has been at the expense of service and supply occupations and of combat skills.

This means that, over the years, a large proportion of military positions—even in the Army and the Marine Corps—came to involve skills requiring aptitudes that, by traditional armed forces standards, are found to a far greater extent among white than among black youths.[25] As proportionately more blacks now populate an Army with proportionately fewer occupations for which they are likely to qualify, they are greatly overrepresented in these occupations, which include artillery, supply, and food service (as shown in table 3-7).

The extent to which the military's needs for highly trained technicians will continue to expand is uncertain. Some factors could increase the requirements for skilled personnel; others could have the opposite effect. Among the former is the introduction of new technologies, some of which are still on the drawing board. Without major breakthroughs,

25. Whether these standards are valid is discussed in chapter 5.

however, there are unlikely to be more than modest changes in the skill composition of the armed forces. Despite the popularized image of the automated Buck Rogers–style battlefield of the future, characterized by small numbers of highly trained operators remotely commanding electronic tanks and laser death rays, the composition of U.S. ground combat forces will probably change little over the next twenty-five years, if past experience is any guide. Although major advances are expected in precision-guided munitions and perhaps in improved battlefield mobility, the demise of traditional ground formations and their heavy dependence on the combat infantryman seems unlikely.

Although the future role of the Navy is far from set, the composition of the fleet in the year 2000 is fairly well determined, given the long lead time associated with shipbuilding programs and the long life expectancies of these expensive naval systems. The Navy's need for skilled personnel is certain to increase in absolute terms if only because of the projected increase in the number of naval vessels. Moreover, the roster of skills is likely to become even more heavily dominated by specialists and technicians since the new ships that will enter the fleet over the next two decades are bound to have more sophisticated electronic systems and more complex propulsion plants than are installed in the older ships being replaced.[26] The change in the Navy's skill mix could be sizable, particularly if more reliance is placed on nuclear propulsion.

The Air Force's roster of skills, already dominated by technicians, is unlikely to undergo more than modest changes over the next two decades. Aircraft and missile systems now in the research and development stage by and large represent improvements in the state of the art rather than revolutionary change. For example, as matters stand, some version of a manned bomber augmented by wide-bodied jets armed with cruise missiles will probably constitute the manned portion of the strategic triad at the turn of the century. At that time the generation of fighter aircraft to follow the F-15/F-16 is likely to be operational. There is nothing now to indicate changes in the design, deployment, or organization of tactical air forces sufficient to warrant modifications in the

---

26. One crude gauge of the growing complexity of naval vessels is shipboard "electrical generating capacity," which provides a proxy for technological sophistication. Between 1964 and 1978 the average electrical generating capacity per ship increased more than twofold. See Barry M. Blechman and others, *The Soviet Military Buildup and U.S. Defense Spending* (Brookings Institution, 1977), p. 50.

manpower skill mix, although one development might lead to major changes in the mix—an expansion of the military's role in space.

Working against further increases in technical skill requirements is the prospect that technological developments will reduce the maintenance work load. This is based on the premise that advances in weapon system design will reduce the need for highly skilled electronic technicians capable of dismantling and repairing sophisticated equipment and instead make possible the maximum use of modern components that can be removed, replaced, and returned for repair to civilian-manned depots.

Thus projecting military occupational requirements is subject to a wide variety of unpredictable influences. It seems safe to conclude, however, that the military's requirements for skilled technicians will continue to increase for the remainder of the century, if more slowly than they have since World War II. The increase will place upward pressure on enlistment standards, and unless current measurement devices are altered or the technical aptitudes of young blacks improve, relatively fewer will qualify for military service.

The Prospective Military Buildup

The military opportunities for blacks may also be affected by the Reagan administration's blueprint for a buildup of U.S. conventional forces. Although the details of the defense program beyond 1983 have not been divulged, Secretary of Defense Caspar Weinberger acknowledged the possibility of further increases in military strength—of as much as 250,000—by the middle of the decade.[27] Apparently, these increases would support the planned buildup to a 600-ship Navy and would fill out the capabilities of air and ground components of the so-called rapid deployment forces.

An expansion of this magnitude, which would substantially increase the annual requirement for new recruits, would allow an increased intake of white volunteers *without* restricting the opportunities of blacks. But whether the armed forces will be able to attract the additional recruits is an open question. What is clear, however, is the gathering consensus that a range of policy options should be exercised to improve the quality and, by implication, the representativeness of the nation's military establishment.

27. *San Diego Union,* May 31, 1981.

Military Manpower Policies

Left to itself, the growth in the proportion of minorities in the enlisted ranks of the nation's ground forces, reinforced by the demographics and economics of the 1980s, would continue. But steps taken in the early 1980s to improve the "quality" of military manpower will counteract the growth. Already established, for example, are incentives designed to attract more high school graduates with above-average aptitudes for military skills who, advocates of a representative force hope, will be from white middle-class America. Also, Congress has legislated qualitative criteria designed to restrict the intake of "subpar" soldiers. Should these programs fail to bring the armed forces up to appropriate standards, administration and congressional leaders have indicated that the all-volunteer system may come to an end. If it does, the nation might adopt a narrow selective service system, a broad national service program, or something in between. In any event, the future racial composition of the armed forces could hang in the balance.

Recruitment Standards

Since the end of conscription in 1973, a central question has been whether the volunteers manning the armed forces are up to the task of protecting vital national security interests. Of major concern has been a deterioration in the quality of recruits as measured by level of education and standardized test scores. The concern deepened when the Pentagon divulged that recruits' test scores reported between 1976 and 1980—already considered extremely low by many observers—had been appreciably overstated. Alarmed by these developments, in 1980 Congress enacted the following qualitative recruitment constraints:

for fiscal year 1981, the Army can enlist or induct no more than 35 percent non-high school graduates among the male non-prior service accessions in the Army;

for fiscal year 1981, the services *as a whole* may not enlist or induct more than 25 percent of the new accessions from personnel who score between the 10th and 30th percentile on the entrance examination;

for fiscal year 1982 *each* service may not enlist or induct more than 25 percent of its new accessions from personnel who score between the 10th and 30th percentile on the entrance examination;

for fiscal 1983 and each fiscal year thereafter, each service may not enlist or induct more than 20 percent of its new accessions from personnel who score between the 10th and 30th percentile on the entrance examination.[28]

As standards are raised, whether by Congress or by the services themselves, fewer blacks will be eligible to enlist and the proportion of black enlistees will drop, provided enough white volunteers can be found. It was already evident by 1981 that the military services, particularly the Army, were meeting the congressional goals and in so doing were reducing the intake, in both relative and absolute terms, of blacks. Effective January 1, 1982, the Army also raised its standards for reenlistment. Under new rules, reenlistment would be denied to soldiers who fail to make grade E-4 (corporal or specialist four) during their first three years of service, first-term soldiers who have accrued two or more AWOLs or one AWOL in excess of five days during their previous twenty-four months of service, and first-term soldiers who are in grade E-4 or below who had failed to score at least 95 on any three components of the standardized aptitude tests used for classification.[29] This policy change, apparently made possible by an improved retention picture, will effectively deny reenlistment to a large proportion of the soldiers who failed to meet minimum standards but were inadvertently recruited

28. *Department of Defense Authorization Act, 1981*, Conference Report, Title III, sec. 302. The restriction on nongraduates was extended into fiscal 1982; *Department of Defense Authorization Act, 1982*, Conference Report, H. Rept. 97-311, November 5, 1981. While this marks one of the few intrusions by Congress into military qualitative standards, the individual services have frequently varied the standards to respond to conditions of supply and demand. This control has been mainly in the form of minimum required scores on supplementary tests. Since 1958, for example, the Army has required that recruits who attain a score on the standardized entry test above the minimum requirement but below the thirty-first percentile (category IV) must also attain certain scores on the so-called AQB (Army Qualification Battery), which was designed to measure specific occupational aptitudes. Qualification standards have been radically changed from time to time, depending on need, as the following vividly illustrates:

Period	Percent in category IV who qualified under AQB requirements
Before August 1958	100.0
August 1958–May 1963	68.1
May 1963–November 1965	31.2
November 1965–April 1966	42.9
October 1966–December 1966	72.0
December 1966	85.1

Bernard D. Karpinos, *AFQT: Historical Data (1958–1972)*, Special Report ED-75-12 (Alexandria, Va.: Human Resources Research Organization, 1975), p. 21.

29. Larry Carney, "Re-Up Bar Tightened for 18-Year Troops," *Army Times*, January 4, 1982.

between 1976 and 1980 (as discussed in chapter 5). Moreover, intended or not, the tighter criteria will probably have a disproportionate impact on blacks, who in recent years have been reenlisting at about 1.6 times the rate for whites. The success in attracting and retaining volunteers with better credentials is attributable at least in part to recent increases in military pay and benefits.

Pay Increases

In 1980 Congress enacted substantial increases in pay and benefits intended to shore up inadequate military wages and to spur recruitment of higher-quality youth. Among them were an 11.7 percent across-the-board increase in military basic pay and allowances and substantial increases in the number and size of enlistment and reenlistment bonuses; the net increase in average military compensation was some 17 percent in that year. Combined with the 14.3 percent increase granted in 1981, the pay of members of the armed forces was raised by roughly one-third in just two years.

It is difficult to predict the number, much less the characteristics, of individuals who are attracted by increases in financial incentives; so many factors enter into enlistment decisions that no single one can be isolated and assessed precisely. Even when all but the pay factor are ignored, there is less than unanimous agreement.[30] Analysts generally agree, however, that a percentage change in the *ratio* of military to civilian pay would yield an equal percentage change in enlistment rates.[31] In other words, an increase in military pay of 10 percent above the increase in comparable civilian pay would be expected to result in a 10 percent increase in the number of recruits who have a high school diploma and at least an average aptitude for military service. Substantial variation has been found among the individual services, the heaviest

---

30. All estimates should be used cautiously because of the uncertainties and inadequacies inherent in estimation processes. Important variables are often neglected, measurement errors are made, and the mathematical models do not represent the true relationships. It is somewhat reassuring when several investigators using different models and data come up with similar estimates, but the statistical confidence attached to the estimates should be given due consideration. Finally, any projections beyond the range of the sample data should be viewed with skepticism, particularly as they apply to racial differences.

31. In the jargon of economics, the pay elasticity is 1.0. Congressional Budget Office, *Costs of Manning the Active-Duty Military*, p. 85.

responses to changes in pay being estimated for Army volunteers.[32] Differences in probable response are also thought to exist between qualitative groupings. The Congressional Budget Office assumed, for example, that a raise in pay of 10 percent would yield a 10 percent increase in the enlistment of high school graduates who do not plan to go to college, a 7.5 percent increase of those bound for two-year colleges, and a 5 percent increase of those bound for four-year colleges.[33]

The historical record also indicates that responses to changes in pay differ among racial groups. For the most part, studies have concluded that black volunteers have been more responsive than nonblacks to changes in pay levels, but the magnitude of the difference can be only roughly approximated. An analysis of recruitment experience in the early 1970s indicated that a 10 percent raise in military pay relative to civilian pay would increase by about one-third the propensity of black high school graduates to join the Army, though the propensity of all high school graduates would increase just 6.5 percent.[34] A 1978 analysis of volunteer enlistments found that the pay effect on Army nonwhite enlistments was over three times as large as on Army white enlistments.[35]

In evaluating the comparative effects of pay increases on future enlistments, account should be taken of the relative proportions of those qualified who are attracted at current pay levels. Overall, as the 1980s began, the services were recruiting only about 12 percent of the pool of qualified male high school graduates aged eighteen to twenty-three, but the differences by race were striking. As table 4-1 showed, the armed services were attracting about 34 percent of qualified black high school graduates in the relevant age group and only 10 percent of the whites, the implication being that a greater response from whites should be expected for a given increase in pay, other things being equal. Moreover, with a constant need for recruits with high school diplomas, the services

---

32. Daniel Huck and others, "Sustaining Volunteer Enlistments in the Decade Ahead: The Effects of Declining Population and Unemployment," Draft Final Report prepared for the Assistant Secretary of Defense for Manpower, Reserve Affairs, and Logistics (McLean, Va.: General Research Corp., 1978), p. 1-7.

33. Congressional Budget Office, *Costs of the National Service Act (H.R. 2206): A Technical Analysis*, Staff Working Paper (CBO, 1980), p. 30.

34. D. W. Grissmer and others, "An Econometric Analysis of Volunteer Enlistments by Service and Cost Effectiveness Comparison of Service Incentive Programs," Policy Analysis Study 15401 (General Research Corp., 1974), pp. 87–90.

35. Huck and others, "Sustaining Volunteer Enlistments," p. 1-8.

Table 6-6. Army Recruits Who Enlisted under Bonus Program, by Race, 1976, 1978, and 1980

| | 1976 | | | 1978 | | | 1980 | | |
| | | Recruits enlisting under bonus program | | | Recruits enlisting under bonus program | | | Recruits enlisting under bonus program | |
Racial group	Total recruits	Number	Percent of total	Total recruits	Number	Percent of total	Total recruits	Number	Percent of total
White	130,838	9,760	7.5	75,863	9,822	13.0	101,591	11,049	10.9
Black	43,520	3,037	7.0	41,863	5,491	13.1	45,565	4,437	9.5
Other	4,521	282	6.2	4,515	407	9.0	8,729	595	6.8
Total	178,897	13,079	7.3	122,141	15,720	12.9	156,885	16,081	10.3

Source: Derived from data provided by Defense Manpower Data Center, April 1981.

would be likely first to accept graduates with the highest scores on entry tests (who, to judge from the past, would be more likely to be white), so the net effect would be a reduction in the proportion of blacks. These factors help explain why the number and proportion of black Army recruits declined in fiscal years 1980 and 1981.

A related point deserves emphasis. More and more, the wisdom of granting pay increases across the board is being challenged as arguments for selective increases to cope with specific problems are advanced.[36] One type of selective pay—enlistment bonuses—has generally been confined to those volunteering for a combat skill, and eligibility has been restricted to high school graduates who score above the thirtieth percentile on the standardized entry test. If these bonuses have had a different effect on the propensity of whites and blacks to volunteer, it is not discernible from recent experience. As table 6-6 indicates, roughly the same proportions of black and white volunteers received bonuses (though members of other racial or ethnic groups lagged behind both).[37] But this does not mean that the racial mix of the armed forces would be insensitive to the extent to which future pay increases were selectively targeted. So far these bonuses have been used almost exclusively for combat positions, which do not generally require high aptitudes for technical and specialized skills. If greater emphasis is placed on targeting pay increases to higher-skill specialties, under current standards relatively fewer black youths will qualify.

Educational Benefits

Some critics of the socioeconomic composition of the armed forces oppose pay increases of any form on the ground that financial inducements, at almost any feasible level, would not motivate middle-class youths to volunteer for the armed forces. Besides, many of the problems of the all-volunteer force have been attributed to "a redefinition of military service in terms of the economic marketplace and the cash-work nexus."[38] These critics contend that educational benefits would be far

36. See Martin Binkin and Irene Kyriakopoulos, *Paying the Modern Military* (Brookings Institution, 1981).
37. These data do not appear to support the contention of a former top Pentagon equal opportunity official. In 1979 Deputy Assistant Secretary for Equal Opportunity M. Kathleen Carpenter told a meeting of the NAACP that black recruits may not be getting their fair share of bonus money. *Army Times*, September 24, 1979.
38. Charles C. Moskos, "Education as Inducement to All-Volunteer Military," *Washington Star*, April 5, 1981.

more effective than cash pay to attract middle-class, college-bound youth and enable the nation to field armed forces more representative of society.

Various legislative proposals to increase the type and level of educational benefits were introduced in the Ninety-seventh Congress in 1981. Most would resurrect the GI Bill in some new form to replace the Post-Vietnam Era Veterans' Educational Assistance Program (VEAP), which has been in effect since January 1977. Under the basic VEAP, service members voluntarily contribute to an education fund. The maximum contribution by service members is $2,700, which, matched on a two-for-one basis with $5,400 of Veterans Administration funds, provides a total of $8,100 for the veteran's educational expenses. The secretary of defense is authorized to provide additional VEAP incentives (called "kickers") for recruiting and retention purposes. Initially, certain participants could receive up to $6,000 in additional funds.[39] Subsequent changes were incorporated in 1980 to raise the maximum amount of educational benefits to $20,100, and in December 1980 several more regional programs were initiated to test the effects of (1) providing a cash payment for educational costs on completion of service, much like the old GI Bill; (2) having the government pay all VEAP contributions; and (3) having the government pay off outstanding federal student loans for those who enlisted in the active Army or the reserves. In fiscal 1982 the Army extended eligibility for the supplemental VEAP kickers to "high-quality" recruits enlisting in seventy-two occupations nationwide; the other services continued to offer only basic VEAP without the added bonus.

The participation rate in VEAP has been disappointing. During the first three years of the program, less than one of every four enlisted personnel opened a VEAP savings account, with the highest rate occurring in the Army (30.2 percent).[40] And minorities were more apt to

39. Chapter 2, Title 38, U.S.C. In January 1982 the Department of Defense assumed financial responsibility for the two-for-one matching benefits as well as the "kickers." The Veterans Administration still retains administrative responsibility for the program.

40. *Third Annual Report to the Congress on the Post-Vietnam Era Veterans' Educational Assistance Program* (Office of the Assistant Secretary of Defense for Manpower, Reserve Affairs, and Logistics, 1980), p. 2-7. The "participation rates" do not account for those who either suspend active participation or drop out of the program entirely. A survey administered in 1979 revealed "considerable movement by eligible enlisted personnel into, out of, and sometimes back into the program." The survey results also suggested that a person's "inability" to save was the primary reason for dropping out; the program barred or discouraged participation by persons with competing financial responsibilities, and

participate than whites, as shown below (as a percentage of those eligible from January 1977 through December 1979):[41]

Racial-ethnic group	Army	Navy	Marine Corps	Air Force	All services
White	27.1	25.9	17.9	7.5	20.8
Hispanic	40.4	34.5	20.6	9.4	31.6
Black	32.8	33.6	19.3	5.5	27.6
Other	44.0	40.1	23.0	10.8	34.4
All groups	30.2	27.7	18.5	7.3	23.3

The differences would probably have been more pronounced if account had been taken of the fact that white participation rates are larger than they would be without the "kicker," since these supplementary inducements have been offered exclusively to high school graduates who score at or above the fiftieth percentile on the standardized entry test, the majority of whom are white.

There is little evidence that returning to the Vietnam era type of GI Bill would attract more highly qualified whites. It has been argued that the chances are slim as long as the system of educational benefits available to students through other government programs remains intact. In 1981 federal aid to college students amounted to about $11 billion in the following programs: Basic Educational Opportunity Grants; Supplemental Educational Opportunity Grants; College Work-Study Program; National Direct Student Loan; and Guaranteed Student Loan. Thus young people who are interested in higher education have had hardly

---

there was "an underlying dissatisfaction with both the administration of the program and the program itself." See Mark J. Eitelberg and John A. Richards, *Survey of Participants and Inactive/Former Participants in the Post-Vietnam Era Veterans' Educational Assistance Program: Results and Conclusions*, FR-ETSD-80-11 (Human Resources Research Organization, 1980). It is estimated that no more than 15 percent of service members who are eligible for VEAP will ever attend school as veterans receiving VEAP benefits. See Congressional Budget Office, *Improving Military Educational Benefits: Effects on Costs, Recruiting, and Retention* (GPO, 1982), p. 15.

41. *Third Annual Report . . . on the Post-Vietnam Era Veterans' Educational Assistance Program*, p. 2-10. The relatively low level of participation by Air Force personnel is attributed in part to the fact that they tend to enlist for a longer period and consequently delay enrollment until they can better afford it. Others have suggested, only partly in jest, that Air Force recruits, who tend to score higher on aptitude tests, recognize a bad deal when they see one. In view of the tendency of the program to discriminate against those with fewer financial resources—those who can least afford to set aside what amounts to a considerable portion of their take-home pay—the relatively high minority participation rates seem even more impressive.

any incentive, so the argument goes, to pursue the military option; and if it is to provide an incentive, a new GI bill would have to be more attractive than the alternatives available.

The Congressional Budget Office, however, disputes that contention. In an analysis of student aid programs, the CBO found that, although these programs have increased access to more costly schools for those who had already made a decision to attend college, they did not appear "to be the major determinant of the basic decision whether or not to attend school."[42]

Several versions of a new GI bill have been proposed. By and large, each would provide a stipend whose size would be tied in some fashion to total time served. At issue, however, is whether the benefits could be used while the member was on active duty, whether they could be converted to cash, and whether they could be transferred to the service member's dependents.

At any rate, there is insufficient empirical evidence to support the contention that the adoption of a GI bill will alter the socioeconomic or racial profile of the armed forces. In fact, skepticism seems warranted by the historical record. In an analysis of the marketplace effects on prospective military volunteers of educational benefits conducted in the 1970s, it was found that young blacks, particularly those showing an interest in the Army, attach greater importance than whites to the GI Bill. Among survey respondents in the National Longitudinal Study of the High School Class of 1972, 57 percent of the blacks who planned to enlist in the military, but only 36 percent of the whites, indicated that the GI Bill was a very important incentive.[43] A more recent (1978) survey of high school youth also suggested that relatively more blacks than whites would be attracted to the armed forces by free educational benefits.[44]

42. "Military Educational Benefits: Proposals to Improve Manning in the Military," Staff Memorandum (Congressional Budget Office, September 1981), p. 14. The Reagan administration's budget for fiscal 1983 contained proposals that, if enacted, would significantly reduce financial aid to students. It is hard to predict how many students would be forced to withdraw from or cancel plans to attend college, how many would shift from full time to part time, how many would transfer from higher-priced to lower-priced colleges, and how many would turn to the military services.

43. Richard L. Eisenman and others, "Educational Benefits Analysis" (Human Resources Research Organization, 1975), p. 27. An interesting finding in this study was that even though blacks attributed greater importance to the GI Bill as an enlistment incentive, they would be more likely to enlist than whites should the GI Bill be terminated.

44. *Youth Attitude Tracking Study* (Market Facts, Inc., 1980).

Peacetime Conscription

Despite a gathering consensus that the nation's armed forces are plagued by manning problems and public opinion polls that show support for a military draft, few believe that the nation will return to peacetime conscription unless a more palpable threat to U.S. security interests develops.[45] Yet Reagan administration officials have more than hinted that if increased pay and benefits fail to upgrade the quality of military manpower a return to some form of conscription will be considered. In that event legislators who have always been skeptical about volunteer forces may well be joined by the ambivalent majority, who might feel that supporting conscription was justified because other approaches had been exhausted. Regardless of rationale, the adoption of compulsory service would be expected to fill the Army's ranks with soldiers from a cross section of American society. The extent to which that would actually occur, however, would depend on the specific form of conscription and on the entry standards imposed on both enlistees and inductees.

SELECTIVE SERVICE. Most advocates of a return to peacetime conscription stop short of proposing a system that would merely meet the needs of the armed forces. The few who do are careful to urge that the problems of the past be avoided. Presumably they would advocate a stochastic, or lottery, system with few exemptions or deferments, which would mean in theory that inductees from each age cohort would represent a cross section of the eligible population. The racial characteristics of the armed forces would then depend in large measure on the mix of volunteers and conscripts, which in turn would depend on the size of the draft calls and on the eligibility standards for both enlistees and inductees.[46]

---

45. A Gallup poll conducted in 1981 indicated that 48 percent of the population thought that the nation should return to a military draft and 45 percent that it should not. *The Gallup Report*, August 1981, p. 15.

46. Other factors would also have to be considered. For one thing, conscription would induce more "volunteers" to enlist in the armed forces. Under the lottery system put into effect in the late 1960s, many of those with low numbers, seeking to avoid duty with the Army in Vietnam, enlisted in the reserves or the active Air Force or Navy. When it became obvious that those services were obtaining the bulk of the high-quality youth, ground rules were devised to encourage each of those services to accept a fair share of recruits who scored in the lower AFQT categories. This explains why the Air Force and the Navy recruited more category IVs during those years than a free-market operation would have required. It also suggests that a return to conscription could mean that the Navy and the Air Force would once again be obliged to accept people with lower aptitudes than would

Table 6-7. Percentage Distribution of Male Army Entrants, Fiscal Year 1981, and Two Conscription Alternatives, by Race, AFQT Category, and Level of Education

Level of education and AFQT category	Fiscal 1981			Alternative 1 (1981 size plus 100,000; volunteers limited to 35% non-high school graduates, 20% category IV; 1981 entry standards)									Alternative 2 (any size; no volunteers; 1981 entry standards)		
				Volunteers			Draftees			Total			Draftees		
	White[a]	Black	Total	White	Black	Total	White	Black	Total	White	Black	Total	White	Black	Total
High school graduates	55	23	78	36	12	48	32	3	35	68	15	83	85	7	92
Category I	2	*	2	2	*	2	2	*	2	4	*	4	6	*	6
Category II	17	1	18	12	1	13	16	*	16	28	1	29	42	1	43
Category III	21	7	28	16	5	21	10	1	11	26	6	32	27	2	29
Category IV	15	15	30	6	6	12	4	2	6	10	8	18	10	4	14
Non-high school graduates	19	3	22	13	2	15	3	*	3	15	2	17	7	1	8
Category I	*	*	*	*	*	*	*	*	*	*	*	*	*	*	*
Category II	3	*	3	2	*	2	1	*	1	3	*	3	2	*	2
Category III	14	2	16	10	2	12	2	*	2	12	2	14	5	1	6
Category IV	1	1	2	0	0	0	0	0	0	0	0	0	0	0	0
Total entrants	74	26	100	49	14	62	35	3	38	83	17	100	92	8	100
Category I	2	*	2	2	*	2	2	*	2	4	*	4	7	*	7
Category II	20	1	21	15	1	16	17	*	17	31	1	32	44	1	45
Category III	36	9	45	26	6	32	12	1	13	38	8	46	31	3	34
Category IV	16	16	32	6	6	12	4	2	5	10	8	18	10	4	14

Source: Fiscal 1981 figures based on data provided by Defense Manpower Data Center. Other figures are authors' estimates based on assumptions listed under each alternative. Figures are rounded.
*Less than 0.5 percent.
a. White includes all racial and ethnic groups other than black.

There are many possibilities and it is risky to generalize, but a rough estimate of the effects can be illustrated. The number of individuals to be drafted would be the difference between total yearly needs for new recruits and the extent to which those needs could be met through voluntary enlistments. Assuming that the Army will continue to attract the same number of male volunteers with the same qualifications that characterized fiscal 1981 recruitment, table 6-7 illustrates the effects on the racial mix of new entrants under two alternatives.

Under the first, the size of the Army is expanded by 100,000, which increases the annual requirement for new recruits to about 135,000, the constraints on proportions of high school dropouts and category IVs established by Congress in 1980 are applied to volunteers, and conscription is used to bring the number up to the required level. In this case, the Army would accept 84,000 volunteers, of whom 65,000 would be high school graduates and 17,000 would score in category IV. The remainder of the requirement for new entrants (51,000) would be filled by a no-exemption, no-deferment lottery draft, which would yield a perfect cross section of youth eligible for military duty under 1981 standards. In contrast to fiscal 1980, when 28 percent of male Army recruits were black, under this alternative only 17 percent would be black.

Alternative 2 assumes that volunteers are not accepted and that total needs are acquired through a lottery draft.[47] The full intake of Army males would then represent the *eligible* population, of which blacks constitute about 8 percent.

While a smaller proportion of those entering the Army would be black under both alternatives, the impact on the overall racial mix in the Army enlisted ranks would ultimately depend on relative rates of early attrition and reenlistment. If the trends discussed in chapter 3 continue, the proportion of blacks in the Army can be expected to decline, though less than the proportion of black recruits.[48]

---

otherwise be the case unless, of course, inductee standards were set higher than volunteer standards.

47. Precluding voluntary enlistments in the armed forces is not as farfetched as it might seem. For most of World War II (December 1942 to September 1945), men were not permitted to volunteer for *enlistment* in the military services. But they could volunteer for *induction*. This stipulation in effect meant that everyone was processed through the Selective Service System and military induction stations, where they were examined, classified, and assigned. Executive Order 9279, *CFR*, Title 3, 1938–1943, p. 1232.

48. Data on reenlistment and attrition appear in appendix tables B-2 and B-5, respectively.

This rough analysis illustrates the effects that conscription could have on the racial mix of Army recruits. But whether such alternatives are even plausible is an open question, in part because of social and political factors and in part because of the mechanics of implementation. Many skeptics doubt that a no-exemption, no-deferment conscription system could actually be installed. Senator Mark O. Hatfield, one of the principal anticonscription spokesmen, argues:

The fact is that the draft is an enemy of the poor and disadvantaged. The popular new phrase "fair draft" is a contradiction of terms. Statistics abound which demonstrate the discriminatory nature of the draft system . . . in 1967 almost 50 percent of minority men were found unfit for service as contrasted to 25 percent of the white male population. Nevertheless, 30.2 percent of the qualified black group were drafted as compared to 18.8 percent of the qualified whites.[49]

Other questions also need to be addressed. How many additional "volunteers" in each racial category would a return to conscription induce? How would physical standards affect the representativeness of inductees? Would a return to conscription, particularly with qualitative restrictions, invite many to seek conscientious objector status, others to drop out of high school, some to misrepresent themselves as alcoholics and drug abusers, and still others to score artificially low on entry tests? Can stricter qualitative standards be justified in the first place, and what would happen if they were tested in the courts? How would the system deal with those who identified themselves as homosexuals? Would women be drafted? What effect, if any, would an Army-only draft have on the racial mix in the other services?

By and large, proposals to reinstitute a draft solely to meet military needs have not been taken seriously because of these difficult questions as well as the fear of reopening old social wounds. This has led many of the critics of the volunteer system to turn to the concept of national service.

NATIONAL SERVICE. A program that would call for all qualified youth to serve the nation in one capacity or another is attractive not only to those interested in filling the ranks of the military, but also to those with the broader interest of improving American society, most notably to those worried about the predicament of America's young. Concern about military recruitment undoubtedly prompted recent discussion of universal service; now the debate has been joined by veterans of the Peace Corps and VISTA programs of the 1960s, who see a new opportunity to rekindle the spirit reflected in President Kennedy's dictum: "Ask not

49. *Washington Star,* April 24, 1981.

what your country can do for you—ask what you can do for your country."[50]

The concept of national service is instinctively appealing, and recent polls indicate that many elements of American society support it.[51] Confusion surrounds the concept, however, owing to a bewildering variety of proposals put forth over the last several decades by different interest groups. Most of these proposals, which vary mainly in terms of scope and participation, can be fitted into four categories.

Voluntary broad-based national service, which would seek, without compulsion, to involve as many young people as possible in their choice of civilian or military service.

Voluntary targeted national service, which would be directed at involving a relatively small, selected segment of youths in service programs that also offer training and remedial assistance to participants.

Compulsory broad-based national service, which would entail mandatory universal registration and require either military or civilian service for all qualified young people.

Compulsory lottery-based national service, which would require randomly selected individuals to perform either civilian or military service.

Voluntary programs, in the main, enjoy wide acceptance but are considered to be of little use for military recruitment. Small national and local VISTA-type programs aimed principally at disadvantaged youth have been available, and expanding this concept would provide civilian opportunities for more jobless young from poor families. The direction and magnitude of the effect on the military's racial mix would depend on the specific nature of the program. Education grants entailing a subsequent military obligation would undoubtedly increase the number of blacks serving in the armed forces. A program of public service jobs, such as those under the Young Adult Conservation Corps or the Youth Community Improvement Projects, while carrying no service obligation,

---

50. Jacqueline Grennan Wexler and Harris Wofford, both of whom were involved in early discussions of plans for national service in the mid-1960s while working together in the Peace Corps, cochaired the Committee for the Study of National Service, established in 1977. The report of the committee, which recommended that the nation "move *toward* universal service by stages and by incentives but without compulsion," was published as *Youth and the Needs of the Nation* (Washington, D.C.: Potomac Institute, 1979), p. 1; emphasis in original.

51. A 1981 Gallup poll, for example, found that 71 percent of the respondents favored a law requiring all young men to give one year of service, military or civilian, to the nation. *The Gallup Report,* June 1981, p. 18.

might, through postjob counseling, increase the propensity of participants to enlist in the military.

Broader-based voluntary programs would be much larger and would not be aimed at any specific sector. In a version proposed by the Committee for the Study of National Service, the nation would phase into a voluntary program that would encourage all young people and provide them with opportunities to serve the nation and the world community full time for one or more years. "Military enlistment," the commission concluded, "should be recognized as a form of National Service, and service should be re-emphasized as the central mission of the military."[52] This should improve the climate for volunteering, according to one side of the argument, and thus help the armed forces attract young people. The other side fears that such a program would compete with the armed forces for the same pool of qualified individuals and could thus have a net adverse effect on military recruitment. It is impossible to know for certain how many young people would volunteer for which programs, much less from what racial or ethnic backgrounds they would come.

Mandatory programs would be more successful in filling the armed forces, but again, predictions concerning the characteristics of the participants are highly uncertain. Under a broad-based system all youth would be required to participate, presumably for one year, in either civilian or military service. If the choice between the two was voluntary, the program would be unlikely to alter the racial mix in the armed forces to any appreciable degree since, as discussed earlier, black youths generally view military service more favorably than whites. More to the point, a survey of high school seniors conducted by the Department of Education in 1980 indicated that, under a compulsory program, blacks would favor military over public service 1.3 to 1 and whites public over military duty 1.4 to 1.[53]

To obtain a military force fairly representative of society, which is presumably one of the main objectives of a national service program, would require either higher entry standards for military than for civilian service or the mandatory assignment of a cross section of youth to the armed forces. This version of national service would share many of the

52. *Youth and the Needs of the Nation*, p. 5.
53. However, 27.6 percent of the black respondents and 30.1 percent of the white indicated that they would try to avoid either option. Department of Education, National Center for Education Statistics, Bulletin 81-245B, April 10, 1981.

shortcomings of military-only conscription and would also be vulnerable to criticism on several other points. First, the cost of such a program, placed as high as $24 billion a year, is considered prohibitive by many.[54] Second, some worry that 3.5 million youths could not be gainfully employed in government service. Third, a national service draft of this scope might well displace some marginal workers in the labor force. Fourth, compulsory national service might not pass a test of constitutionality.

For all these reasons, several variations of national service programs, generally of narrower scope and hence presumably of wider popularity, have been advanced. The National Service Act (H.R. 2206) introduced in 1979 by Representative Paul N. McCloskey, Jr., would give every eligible youth the opportunity to volunteer for one year of civilian service or two years of military service. If too few chose the military, young men would be drafted by lottery to fill the rest of the requirement. Subsistence wages would be paid to all participants, but military personnel would also receive postservice educational benefits, perhaps similar to those provided during the Vietnam era.[55] Here, again, given the uncertainty about the response of youth to national service, it is difficult to predict how many might volunteer for one or the other form of service. But in light of the proclivity of black youth for military service in general and the attractiveness of educational benefits in particular, the racial composition could remain virtually unaffected unless higher quality constraints or the channeling of selected youth to military service were instituted.

54. Congressional Budget Office, *National Service Programs and Their Effects on Military Manpower and Civilian Youth Problems* (GPO, 1978), p. xvii.
55. Many aspects of H.R. 2206 are vague. For an interpretation of this complex proposal, see Congressional Budget Office, *Costs of the National Service Act;* and Carnegie Council on Policy Studies in Higher Education, *Giving Youth a Better Chance: Options for Education, Work, and Service* (Jossey-Bass, 1979), pp. 275–76.

CHAPTER SEVEN

THE POLICY DILEMMA

It is customary in the democratic countries to deplore expenditures on armaments as conflicting with requirements of social services. There is a tendency to forget that the most important social service a government can do for its people is to keep them alive and free.   —Sir John Slessor

During my seven years in the Department [of Defense] it seemed to me that those vast resources could contribute to the attack on our tormenting social problems . . . in the end, poverty and social injustice may endanger our national security as much as any military threat.   —Robert S. McNamara

CRITICS of the nation's decision to abolish military conscription in 1973 warned that an armed force that raised its manpower solely by voluntary means would become increasingly unrepresentative of the society it was established to protect and defend; of major concern was the prospect of racial imbalances.

These predictions have come to pass, most conspicuously in the Army and to a lesser extent in the Marine Corps. Whereas just before the end of the draft black membership in the enlisted ranks of the nation's ground forces was roughly in line with the eligible population (about 12 percent), by 1981 the proportion had reached just over 33 percent in the Army and 22 percent in the Marine Corps. Growth in the black membership of the Air Force and Navy enlisted ranks was far more modest: by 1981 black representation in the Navy was about equal to and in the Air Force slightly over the black proportion of the military-age population. At the same time, the proportion of black officers in the armed forces remained noticeably out of balance despite a twofold increase over the period, from 2.5 percent in 1972 to just over 5 percent in 1981.

This situation, the result of a variety of social and economic factors not necessarily related to voluntary recruitment, has given rise to a number of worries—some held predominantly by whites, others held predominantly by blacks, and some shared by members of both groups.

The Policy Dilemma 153

Much of the uneasiness may simply be reaction to change from racial proportionality—a situation that is generally understandable and acceptable to all population subgroups. But there are specific worries as well.

Fielding combat forces composed of an overproportion of blacks, some say, imposes an unfair burden on one segment of American society, a burden that seems decidedly inequitable because members of that group have not enjoyed a fair share of the benefits that the state confers. Thus the prospect that as many as half the combat casualties in the early phases of a military engagement would be suffered by black soldiers or marines is immoral, unethical, or somehow runs contrary to the precepts of democratic institutions.

Others believe that military forces containing such a high concentration of blacks pose certain risks to U.S. national security interests. Those who gauge the caliber of recruits by the scores they attain on standardized entry tests point out that four of every ten black volunteers who have entered the armed services since the end of the draft have been in the lowest acceptable mental category, the implication being that they are less trainable, particularly in technical and specialized skills, and that they cause more disciplinary problems. A socially unrepresentative force, others argue, lacks the chemistry needed for the group cohesion considered vital to combat units and hence impairs military unit effectiveness. Still others worry about the reliability of such a force if it were deployed in situations, either domestic or foreign, that would test the allegiance of its minority members. Finally, it is feared that an army composed of such a large proportion of blacks lacks legitimacy, both in the eyes of the American public and from the vantage point of the nation's allies and adversaries.

Largely for these reasons, pressure to make the armed forces more broadly representative of American society has recently intensified. Proposals to that end range from measures to increase financial and educational incentives designed to attract additional white volunteers to calls for some form of conscription or national service that would compel more white youths to serve in the armed forces.

On the other side are those who contend that the mere act of raising the question of representativeness is racist because it implies an inverse relationship between the number of black troops and the overall quality of the force. Besides, they argue, a disproportionately black force is not without its benefits; especially appealing are the employment, training, and social opportunities not otherwise available to many black youths. The *real* issues, according to this view, are the conspicuous underrepresentation of blacks in the officer corps and the continuing racial discrimination in job classification and assignment, promotion opportunity, and the military justice system.

Two separate but related policy issues—one social and the other military—emerge from this complicated array of views, questions, and concerns. The social issue is the trade-off between benefits and burdens discussed in chapter 4. This equity question impales the nation on the horns of a particularly difficult dilemma. Does the fact that blacks will probably die in grossly disproportionate numbers, at least initially, in defense of national interests outweigh the fact that the armed forces provide many blacks with their only bridge from the "permanent underclass" to a better life? That countless blacks have chosen to bear the burden in order to reap the benefit, and in so doing have enabled the nation to maintain a volunteer army, nevertheless begs the question of equity.

But this is a single aspect of a much broader issue: the appropriate concept of military service in contemporary American society. This encompasses far more than racial considerations, for it pits against one another the conflicting philosophies dedicated to protecting individual freedom of choice and to restoring a spirit of service as an important duty of the citizen. At bottom, the question is whether the nation should resurrect the concept of the citizen soldier that was abandoned with the end of conscription or maintain a professional army manned strictly by volunteers.

The principal military issue is the relation between the racial composition of the armed forces and their effectiveness. Speculation, a legacy of racial stereotypes of an earlier era, still surrounds the individual capabilities of blacks, their influence on group performance, and how their participation affects foreign and domestic perception of the U.S. armed forces.

The analysis presented in chapter 5 indicates that, in individual performance, blacks should no longer have to prove themselves. While a substantial portion of black youths do not qualify for military service under present standards, those who do perform at least as well as their white colleagues. Job opportunities for blacks, however, have been disproportionately in the relatively unskilled nontechnical occupations and will remain so as long as current classification standards are unchanged or as long as it takes for blacks to realize the benefits of the programs designed to repair the effects of years of educational and social neglect.

Whether the standards used for enlistment, job classification, and assignment are as valid as the adherence to them implies is an open question. Gauging job performance, much less delineating individual characteristics that foretell performance, is a difficult task. While in many cases present standards are based on years of experience and are the products of extensive and rigorous research, in others they appear to be nothing more than legacies of the conscription era when there was virtually no pressure on the armed forces to justify their manning criteria.

Even if it is assumed that present standards are cost-effective in strictly military terms, there remains the question of whether the nation's military establishment, with its vast array of training courses and modern instructional techniques, should apply these measures in the interest of the social good. There is evidence that a large proportion of those who have entered the armed forces despite their failure to meet established entry standards have become "successful" soldiers after the investment of additional time and resources, but the toll exacted from the defense budget and military readiness has not been measured.

As for group performance, while the association between blacks and whites in the armed forces contrasts favorably with race relations in American society as a whole, less than complete racial harmony has been attained. After all, the military establishment is not isolated from the rest of society nor is it immune to society's problems. Race relations in the United States appear to have entered a new phase as the nation has moved toward the right; a mood of intransigence is evident among many whites, and many blacks seem to prefer social, if not professional, separatism. At the same time, the ground forces are attracting many of their volunteers from the disparate segments of society in which these attitudes are bound to be most pronounced: urban blacks and rural whites. While incidents with racial overtones have increased, it is difficult to judge how widespread and how deep these feelings run or what their consequences will be for military readiness. The implications of the asymmetry in officer and enlisted participation, particularly in the ground forces, are also elusive. Seemingly, an increase in the proportion of black officers ought to improve group relations, but there is no hard evidence to support that. In recent years the armed forces have understandably played down the effects of racial differences, leaving the implications for unit cohesion and performance of the substantial shift in the racial mix virtually unresearched.

Suspicion that black troops might be unwilling to carry out their assignments in certain domestic situations—a suggestion that is understandably reprehensible to many members of the black community—cannot be dismissed out of hand. As the British policy not to send Irish regiments into Northern Ireland during the 1970s illustrates, the deployment of troops that share a racial or ethnic bond with an adversary poses difficulties. Less serious is anxiety about foreign involvements since, with the unlikely exception of an American intervention on the side of whites in a conflict against blacks (for example, to support the South African government), it is difficult to conceive of a situation in which black allegiance would be tested. What is certain is that future adversaries, like most of the nation's previous military opponents, will attempt to precipitate racial divisiveness among U.S. troops.

Public perceptions of the racial composition of U.S. armed forces are difficult to gauge. The Soviet Union faces many of the same questions regarding the racial and ethnic composition of its armed forces; indeed, its problems are probably greater. Because minorities in the Soviet military are considered to be less intelligent or less capable and are relegated largely to unskilled jobs in low-priority units, the Soviets are likely to view a blacker U.S. military as a weaker U.S. military. Moreover, some of America's allies have been increasingly critical of the quality of U.S. troops and the racial overtones in their criticisms are unmistakable. But perceptions at home may be even more serious. Although the evidence is thin, speculation that the image of a "black Army" has cost the nation's military the support of the white middle class cannot be completely discounted.

Separate from fears about disproportionate black casualties and defections of black soldiers in certain contingencies is worry that the mere prospect of the social and military consequences could influence a decision about the use of military force. If that were to happen, the range of choices available to national leaders would be reduced and national security could be diminished.

Given the complexity of these questions and the controversy and emotion they arouse, it is tempting to accept the view that the racial composition of the armed forces is *not* an issue. The nation could continue as it is, ignoring the questions in the hope that they will disappear. But the analysis in chapter 6 suggests that the issues are bound to persist and perhaps intensify over the remainder of the decade as the number—if not the technical abilities—of young Americans declines and the requirement for better—if not more—military recruits increases.

The racial connotations will be significant. Blacks will constitute an increasing proportion of the smaller supply pool and, by our economic reckoning, will find military service even more inviting than in the past. At the same time, the services will be under pressure from Congress and from the demands imposed by advancing technology to seek more youths with higher technical and mechanical aptitudes, who by present measuring sticks are more likely to be white.

Manning the armed forces with a representative cross section of American society would be no mean task. As long as the forces are maintained by voluntary means, it will prove difficult, even with large pay increases and more liberal benefits, to attract white middle-class youths, particularly those who are college-bound. Peacetime conscription would make the quest easier, but only if qualitative standards were set high enough to restrict the number of volunteers or if the size of the armed forces was increased. Moreover, a national service program, in which most if not all young Americans participated, would have to either encourage white youths to choose military over civilian service or assign everyone in accordance with representational quotas.

It is important that the nation address head-on the questions raised in this study and evaluate the social and national security choices they entail. The debate should not be confined to racial considerations but should encompass the broader issue of *the role of military service in contemporary American society*. That issue was last raised at the height of the Vietnam War by a commission appointed by President Nixon, whose report led to the abandonment of military conscription. The need to examine the desirability of maintaining that course has become increasingly apparent since the volunteer system was adopted. The racial issues discussed in this study emphasize the urgency of that need.

That so many Americans are ambivalent about military service is understandable; the arguments on both sides are powerful, and honest differences of opinion are difficult to reconcile. Resolution of the issue will be a formidable task requiring study, negotiation, and compromise. Two steps should be taken in preparation for that task.

First, research should be undertaken designed to reduce the uncertainty surrounding the questions raised in this study. The research should cover the following four areas.

Appropriate measures of individual job performance and standards of eligibility need to be developed to better define the relationship of test scores and educational attainment to job performance. This research would be particularly important if, *on the basis of stricter standards*, blacks were squeezed out of the volunteer military or whites were forced into a conscripted military.

The costs—both budgetary and in military readiness—of accepting and training individuals who fall below "normal" standards should be evaluated, as should the benefits of doing so, in terms of enrichment of human capital. These data are essential to any assessment of the cost-effectiveness of existing standards or of a broader social role for the military establishment.

Studies should be made to determine the relationship between racial mix and group cohesion and performance. Difficult though it is to broach this question, it is imprudent to assume that race relations in the armed forces no longer need consideration.

Further attention should be directed toward the enlisted-officer imbalance. Attracting more black officers will prove difficult so long as a college degree is considered an important criterion. The extent to which commissioning programs—such as officer candidate schools, academy preparatory schools, and direct appointments, which are already increasing the opportunities for black non-college graduates—can be expanded to provide additional black officers must be evaluated.

Second, the president should appoint an advisory commission on military service. Unlike its predecessor of the Vietnam era, which in effect was chartered to develop a plan for ending conscription, and unlike the interagency Military Manpower Task Force, which was established by President Reagan in 1981 to study ways to preserve the all-volunteer force, this commission should be charged with examining *alternative* concepts of service to the nation, both military and civilian, that would contribute to the national security, economic, and social environments of the 1980s. The commission should be made up of eminent citizens of multidisciplinary backgrounds and representative viewpoints. The report of this commission could provide a basis for long-overdue public consideration of the issues.

As THE UNITED STATES comes to grips, as it soon must, with the gathering problems of manning its military establishment, on the one hand, and of upgrading the status of its "underclass," on the other, the policy interactions cannot be ignored. To do so runs the risk that national security decisions will be made at the expense of the social good and social decisions at the expense of national security, with a good chance that both will suffer. We hope that this study, which is but a first step in promoting a better public understanding of the issues involved, stimulates interest in developing national policies that will lead to both a stronger military establishment and a healthier society.

APPENDIX A

Racial-Ethnic Categories in the Armed Forces

RACIAL and ethnic standard classifications are established for governmentwide record-keeping and the collection and presentation of data on race and ethnicity. The classifications, the government points out, "should not be interpreted as being scientific or anthropological in nature, nor should they be viewed as determinants of eligibility for participation in any Federal program."[1]

This caveat is not misplaced. There is widespread disagreement among biologists, anthropologists, psychologists, sociologists, and others over the nature and use of classifications of "race." The term is applied in many contexts; it is defined both formally and informally in numerous ways; and it has at least four common usages: (1) biological or physical anthropological; (2) mystical or "romantic"; (3) formal-legal or administrative; and (4) social.

Because of confusion, ambiguity, muddled definitions, and a history of misuse, some scientists advocate abandoning the term (and associated concepts) in favor of "ethnic groups" or some other "noncommittal" phrase.[2] Others believe that substitution is impossible since "race" is so much a part of the scientific literature and the language of our society. "Race is an explosive term," Brewton Berry and Henry L. Tischler wrote. "Our language does have its full quota of 'loaded' words . . . but when it comes to arousing people's prejudices, loyalties, animosities, and fears, none is the equal of race." They added, "However, we can bear in mind that race has both a biological and a social meaning, and that it is the latter that takes precedence in the affairs and thinking of

1. Directive No. 15, "Race and Ethnic Standards for Federal Statistics and Administrative Reporting," *Federal Register*, May 4, 1978, p. 19269.
2. See Ashley Montagu, *Man's Most Dangerous Myth: The Fallacy of Race*, 4th ed. (World, 1964), pp. 372–80.

most of us."[3] The *International Encyclopedia of the Social Sciences* similarly observes that "race engages the attention of social scientists as a special constellation of cognitive or ideological categories and as a means of explaining sociocultural phenomena." The "evaluation of the relevance of racial differences to sociocultural theory," it points out, "thus becomes an inescapable obligation of the social sciences."[4]

It is far beyond the scope of this paper to discuss the concept of race. It is important to note here only that concepts and typologies based on racial distinctions form a major part of the literature on military membership and civil-military relations. Studies of racial group differences in this country most frequently deal with the white and black "races" (because of relative population sizes, American history, and the sociopolitical milieu).

The Department of Defense uses a combined format for racial and ethnic data. The racial categories are Caucasian, Negro, Asian or Pacific Islander, American Indian or Alaskan Native, and "unknown" for those who do not identify their racial heritage as one of the above or who neglect to indicate racial heritage. The ethnic categories are Mexican-American, Puerto Rican, Cuban-American, Spanish descent, American Indian, Filipino, Chinese, Japanese, Korean, Asian-American, Eskimo, Aleut, other, and unknown.

Table A-1 shows the distribution of military personnel by racial-ethnic groupings. The corresponding percentage distribution of armed forces personnel and of the U.S. population appears in table A-2.

3. Berry and Tischler, *Race and Ethnic Relations,* 4th ed. (Houghton Mifflin, 1978), pp. 23, 42.
4. *International Encyclopedia of the Social Sciences* (Macmillan and Free Press, 1968), vol. 13, p. 263.

Table A-1. Military Personnel by Race or Ethnic Category, September 1981
Thousands

Race or ethnic group	Army		Navy		Marine Corps		Air Force		All services		
	Officers	Enlisted	Officers	Enlisted	Officers	Enlisted	Officers	Enlisted	Officers	Enlisted	Total
White	88.8	394.3	61.2	369.6	17.3	120.7	90.8	354.6	258.1	1,239.2	1,497.3
Black	7.9	223.8	1.8	56.1	0.7	37.9	4.8	77.0	15.2	394.8	410.0
Hispanic[a]	1.2	30.1	0.5	14.5	0.2	9.5	1.7	18.3	3.5	72.4	75.9
Asian/Pacific Islander	0.8	7.9	0.7	21.7	0.1	1.3	1.3	7.4	2.8	38.3	41.1
American Indian/ Alaskan Native	0.2	2.7	0.2	3.2	*	1.2	0.6	6.5	1.0	13.6	14.6
Other or unknown	2.9	15.9	1.3	4.1	*	1.8	0.5	3.3	4.8	25.1	29.9
Total	101.8	674.7	65.7	469.1	18.4	172.3	99.6	467.2	285.5	1,783.4	2,068.9

Source: Data provided by Department of Defense, Defense Manpower Data Center.
*Fewer than 50.
a. Hispanics may be of any race.

Table A-2. Percentage Distribution by Race or Ethnic Category, Military Personnel, September 1981, and Total Population, 1980

Race or ethnic group	Army		Navy		Marine Corps		Air Force		All services		U.S. population
	Officers	Enlisted	Officers	Enlisted	Officers	Enlisted	Officers	Enlisted	Officers	Enlisted	
White	87.2	58.4	93.2	78.8	94.1	70.1	91.1	75.9	90.4	69.5	79.6
Black	7.8	33.2	2.7	12.0	4.0	22.0	4.8	16.5	5.3	22.1	11.5
Hispanic[a]	1.1	4.5	0.8	3.1	1.0	5.5	1.7	3.9	1.2	4.1	6.4
Asian/Pacific Islander	0.8	1.2	1.0	4.6	0.5	0.7	1.3	1.6	1.0	2.1	1.5
American Indian/ Alaskan Native	0.2	0.4	0.3	0.7	0.2	0.7	0.6	1.4	0.4	0.8	0.6
Other or unknown	2.9	2.4	2.0	0.9	0.2	1.1	0.5	0.7	1.7	1.4	0.4

Sources: Military personnel based on data provided by Defense Manpower Data Center, May 1981. U.S. population figures derived from unpublished 1980 census data provided by Department of Commerce. Figures are rounded.
a. Hispanics may be of any race.

APPENDIX B

Statistical Tables

Table B-1. Blacks as a Percentage of Selected Reserve Forces, by Component, Fiscal Years 1972–81

Fiscal year	Army National Guard	Army reserve	Naval reserve	Marine Corps reserve	Air National Guard	Air Force reserve	Total selected reserves
1972	2.0	2.9	3.0	7.4	1.4	3.3	2.6
1973	3.2	5.6	3.5	12.6	2.0	4.2	4.2
1974	5.6	7.2	3.4	11.6	2.9	5.6	5.6
1975	7.2	11.1	4.4	14.1	3.8	8.1	7.8
1976	10.6	14.8	5.4	15.4	4.8	9.7	10.5
1977	14.5	19.6	5.9	18.0	5.7	11.8	13.8
1978	16.5	21.6	5.9	19.3	6.4	13.2	15.4
1979	16.9	23.3	6.7	20.1	6.8	14.0	16.0
1980	16.7	23.6	7.1	19.9	7.1	14.3	16.3
1981	16.6	23.9	7.9	19.8	7.3	14.7	16.6

Sources: *Hearings on Military Posture and H.R. 5068: Department of Defense Authorization for Appropriations for FY 1978*, Hearings before the House Committee on Armed Services, 95 Cong. 1 sess. (Government Printing Office, 1977), p. 1187; Robert L. Goldich, "Military Manpower Policy and the All-Volunteer Force," Issue Brief IB77032 (Congressional Research Service, December 1980); and Deputy Assistant Secretary of Defense for Reserve Affairs, "Official Guard and Reserve Manpower Strengths and Statistics" (September 1981), p. 61.

Table B-2. Army Reenlistment Rates, by Race and Career Status, and Racial Composition of All Army Reenlistments, Fiscal Years 1972–81
Percent

Fiscal year	Army reenlistment rates[a]				Racial composition of Army reenlistments	
	First-term		Career			
	White	Black	White	Black	White	Black
1972	12.6[b]	20.4[b]	42.6	61.3	79.8	18.8
1973	35.7[b]	46.1[b]	60.9	69.8	78.1	19.9
1974	26.6	43.3	70.4	80.5	77.6	20.9
1975	33.4	54.1	70.3	82.7	74.9	23.5
1976	29.4	42.2	69.1	82.0	71.8	25.9
1977	30.5	48.4	66.3	80.3	70.5	27.7
1978	27.8	47.5	63.4	78.0	68.7	28.7
1979	33.5	53.7	59.6	74.9	63.4	33.4
1980	45.1	65.1	66.3	79.6	60.2	36.1
1981	44.9	66.4	68.0	81.9	57.9	37.5

Source: Data provided by the Department of the Army.

a. Reenlistment rates for first-term and career-eligible persons who are considered qualified and in specified categories for reenlistment are statistically adjusted to include only those scheduled to separate from active duty during the fiscal year.

b. Reenlistment rates in 1972 and 1973 are for persons who originally entered the Army as volunteers. In 1972 the reenlistment rates for white and black draftees were 11.8 percent and 14.8 percent, respectively. In 1973, 10.6 percent of all eligible first-term white draftees and 12.4 percent of all eligible first-term black draftees reenlisted.

Table B-3. Percentage of New Recruits with a High School Diploma, by Race and Service, 1972–81[a]

Fiscal year	Army		Navy		Marine Corps		Air Force		Total	
	White	Black	White	Black	White	Black	White	Black	White	Black
1972	61.0	63.6	77.8	69.2	46.9	47.0	85.0	80.6	68.1	65.1
1973	62.8	58.7	71.3	66.5	44.0	46.4	84.7	85.8	67.6	63.0
1974	49.3	49.5	68.8	66.8	46.5	45.7	84.9	90.1	61.2	57.3
1975	56.4	60.3	73.4	71.6	54.4	48.2	85.9	90.7	66.3	64.7
1976	55.6	63.6	74.6	80.3	57.8	63.2	88.8	92.0	67.3	68.7
1977	57.2	66.6	72.9	77.5	69.2	75.2	92.0	95.6	70.5	72.2
1978	69.9	78.9	70.8	80.1	68.1	76.4	84.8	91.9	73.6	80.4
1979	60.8	70.6	71.7	84.7	69.6	79.7	82.6	91.0	70.4	76.6
1980	49.4	66.6	72.8	87.2	74.2	85.4	83.2	91.3	66.2	75.2
1981	76.3	90.6	73.7	86.8	77.6	87.3	88.6	94.4	78.8	90.2

Source: Derived from data provided by Defense Manpower Data Center.
a. Recruits with a high school diploma include those who attended or graduated from college. Individuals who passed the General Educational Development high school equivalency examination are not included.

166 Blacks and the Military

Table B-4. Estimated Family Income, by Race, from 1979 Defense Department Survey of Personnel Entering Military Service
Percent unless otherwise specified

Family income[a] (dollars)	Participant responses[b]	
	Whites (4,592)	Blacks (1,710)
Under 2,599	1.8	9.0
2,600–5,199	3.7	10.8
5,200–10,339	14.7	23.3
10,400–15,599	18.0	12.8
15,600–20,779	16.7	7.4
20,800–25,999	11.3	3.3
26,000–31,199	7.5	2.8
31,200 or more	10.4	3.2
No response	15.9	27.4
Total	100.0	100.0

Source: Derived from data provided by Defense Manpower Data Center.

a. Survey question: "What would you say is the total yearly income of your parent(s) or guardian(s) and all family members who lived with them, before taxes and other deductions? DO *NOT* INCLUDE YOUR INCOME. (Give your best estimate. Mark only one answer.)"

b. A brief description of the survey and the survey respondents is presented in table B-14, footnote a.

Table B-5. **Military Personnel Not Completing Their First Enlistment Period, by Education, Race, and Sex, Fiscal Years 1972-78**[a]
Percent

	Year of entry into military service						
Category	1972	1973	1974	1975	1976	1977	1978
Non-high school graduates[b]							
White and other races							
Male	41.5	46.7	51.7	50.9	48.9	47.0	42.2
Female	63.8	60.1	56.8	56.2	54.6	55.7	51.2
Black							
Male	42.4	46.1	50.6	51.4	47.2	42.9	38.1
Female	56.7	42.0	45.2	41.6	37.2	38.0	35.1
Total							
Male	41.7	46.6	51.4	51.0	48.6	46.2	41.4
Female	62.6	57.1	55.0	54.1	52.4	53.5	49.4
High school graduates							
White and other races							
Male	20.4	22.3	25.4	25.2	25.2	23.7	22.6
Female	43.5	39.7	36.2	35.6	37.2	37.2	36.7
Black							
Male	25.2	27.8	29.3	27.9	27.3	24.8	23.1
Female	38.8	32.9	28.0	25.8	27.3	28.6	29.2
Total							
Male	21.0	23.3	26.2	25.7	25.5	23.9	22.7
Female	42.9	38.6	34.8	33.9	35.6	35.8	35.0

Source: Derived from data provided by Defense Manpower Data Center.

a. The attrition rates shown here reflect all discharges that occurred during the first thirty-six months of the first term (other than expiration of term of enlistment).

b. Includes enlisted personnel who passed the General Educational Development high school equivalency examination but did not attend college.

Table B-6. Trends in Disciplinary Incidents, by Service, Race, and Sex, Fiscal Years 1979–81

Number of persons per thousand average monthly strength

Service, type of incident, and sex	1979 Black	1979 White	1980 Black	1980 White	1981 Black	1981 White
Army						
Unauthorized absence						
Male	38.3	29.5	42.4	34.1	37.6	30.3
Female	12.2	15.2	15.2	22.3	13.4	16.7
Designated deserter						
Male	13.8	18.5	15.1	21.1	11.0	18.3
Female	4.5	7.2	5.6	13.1	3.9	11.3
Navy						
Unauthorized absence						
Male	69.9	52.3	68.0	51.4	61.3	47.1
Female	17.0	18.6	19.1	22.0	14.1	17.4
Designated deserter						
Male	25.9	28.3	23.9	25.6	21.2	21.5
Female	3.3	7.4	3.4	10.1	3.5	7.5
Marine Corps						
Unauthorized absence						
Male	78.7	59.8	107.4	71.9	98.0	63.6
Female	24.3	33.2	30.8	27.3	29.4	26.5
Designated deserter						
Male	29.5	35.0	34.0	35.0	29.1	29.1
Female	5.1	14.0	5.4	11.3	4.6	6.6
Air Force						
Unauthorized absence						
Male	7.3	4.9	7.8	5.2	6.2	4.3
Female	5.8	7.0	4.5	7.2	3.0	5.2
Designated deserter						
Male	1.0	0.7	1.4	1.1	0.8	0.9
Female	0.1	1.6	0.5	1.9	0.6	1.1

Source: Data provided by Office of the Assistant Secretary of Defense for Manpower, Reserve Affairs, and Logistics, May 1981.

Table B-7. Recruits Who Received Moral Waivers, by Type of Waiver, Service, and Race, Fiscal Year 1981[a]
Percent

Type of waiver	Army		Navy		Marine Corps		Air Force		All services		
	White	Black	White	Black	White	Black	White	Black	White	Black	Total[b]
Minor traffic offenses	0.3	0.1	0.3	0.1	43.9	22.3	0.1	*	5.7	2.6	5.1
Less than three minor offenses (nontraffic)	0.7	0.2	0.7	0.3	3.6	2.2	0.6	0.3	1.0	0.5	0.9
Three or more minor offenses (nontraffic)	0.4	0.1	0.3	0.2	0.4	0.2	*	*	0.3	0.1	0.2
Other (nonminor) misdemeanors	7.4	4.0	12.4	7.7	6.0	2.9	5.6	3.0	8.3	4.3	7.4
Adult felony	*	*	0.4	0.4	0.5	0.3	0.3	0.3	0.3	0.2	0.2
Juvenile felony	*	*	0.8	0.7	1.0	0.5	0.3	0.3	0.5	0.3	0.4
Drug abuse	*	*	10.9	7.8	0.4	0.2	0.4	0.2	3.5	1.5	3.0
Alcohol abuse	*	*	0.4	0.1	0.3	0.1	0.0	0.0	0.2	*	0.1
Total	8.9	4.5	26.2	17.3	56.1	28.7	7.4	4.1	19.8	9.5	17.3

Source: Derived from data provided by Defense Manpower Data Center. Figures are rounded.
* Less than 0.05 percent.
a. Interservice comparisons should be avoided since differences could reflect different reporting procedures, administrative processes for enlistment, and, in some cases, entry requirements of the individual services.
b. Total includes all racial and ethnic groups.

Table B-8. Distribution of Enlisted Personnel Discharged from the Armed Forces, by Character of Service, Race, and Branch of Service, Fiscal Year 1980[a]

Percent

Character of service	Army		Navy		Marine Corps		Air Force		All services	
	White	Black	White	Black	White	Black	White	Black	White	Black
Honorable	92.01	90.77	91.37	89.84	91.91	89.03	94.91	94.08	92.61	91.07
General	3.32	4.72	5.25	7.03	3.90	6.62	4.46	5.16	4.17	5.21
Other than honorable	4.28	3.73	2.75	2.46	3.23	3.14	0.54	0.61	2.80	3.02
Bad conduct	0.33	0.59	0.62	0.64	0.71	0.82	0.09	0.14	0.37	0.54
Dishonorable	0.06	0.19	0.01	0.03	0.04	0.13	[b]	0.01	0.03	0.14
Unknown	0.00	0.00	0.00	0.00	0.21	0.26	0.00	0.00	0.02	0.02
Totals										
Percent	100.00	100.00	100.00	100.00	100.00	100.00	100.00	100.00	100.00	100.00
Number	157,016	68,847	103,079	11,301	38,165	9,668	107,111	18,055	405,371	107,871

Source: Data provided by Defense Manpower Data Center.
a. Tabulations include enlisted personnel who were "discharged" for purposes of immediate reenlistment and those who entered officer programs.
b. Records show that two white enlistees in the Air Force were discharged by court-martial under dishonorable conditions.

Appendix B   171

Table B-9. Enlisted Separation Rates, by Cause of Separation, Race, and Sex, Fiscal Year 1980[a]
Percent

Cause of separation[b]	White			Black		
	Male	Female	Total	Male	Female	Total
Completion of enlistment	10.93	7.99	10.69	7.32	4.10	7.00
Retirement	2.40	0.27	2.22	1.44	0.23	1.32
Convenience of government						
Pregnancy	...	4.45	0.36	...	3.13	0.32
Marginal performance	1.74	3.73	1.90	2.15	3.00	2.23
Early release program	1.69	2.35	1.75	1.56	1.11	1.52
Other	1.05	2.22	1.14	0.86	1.49	0.92
Misconduct	0.80	0.46	0.78	0.71	0.23	0.66
Discharge in lieu of court-martial	0.78	0.36	0.75	0.80	0.18	0.74
Unsuitability/apathy	0.56	0.44	0.55	0.66	0.32	0.62
Physical disability	0.37	0.46	0.38	0.34	0.31	0.34
Unsuitability/personality disorder	0.32	0.66	0.35	0.18	0.18	0.18
Dependency or hardship	0.29	0.69	0.32	0.11	0.45	0.15
Personal abuse of drugs	0.19	0.09	0.18	0.21	0.04	0.19
Court-martial	0.14	0.01	0.13	0.21	0.01	0.19
Death	0.12	0.06	0.12	0.11	0.03	0.10
Total[b]	21.51	24.44	21.75	16.75	14.88	16.55
Adjusted total[c]	8.19	16.17	8.83	7.99	10.55	8.24

Source: Derived from data provided by Defense Manpower Data Center. Figures are rounded.

a. The separation rate is the percentage of total enlisted personnel in each category of race and sex (based on fiscal 1980 end-strengths) who were separated (discharged) from active duty for the reason cited. All separations (those that occurred during the final days of fiscal 1980) were not processed by the military services until after the end of the fiscal year and do not appear in the table.

b. Does not include temporary separations (for processing of immediate reenlistments) and separations of enlisted personnel who entered officer programs.

c. Excludes separations resulting from completion of enlistment and retirement.

Table B-10. Blacks as a Percentage of Male Enlisted Personnel Assigned to Occupational Areas, by Service, Selected Years, 1964-81[a]

Occupational category	1964[b]			1972				1976				1981			
	Navy	Marine Corps	Air Force	Navy	Marine Corps	Air Force	Navy	Marine Corps	Air Force		Navy	Marine Corps	Air Force			
Infantry, gun crews, and seamanship specialists	4.1	11.6	2.5	11.0	20.1	4.8	11.3	22.7	6.2	13.8	26.0	15.3
Electronic equipment repairmen	1.9	2.3	4.2	2.4	4.1	4.7	3.3	4.7	5.6	5.3	7.6	7.2
Communications and intelligence specialists	3.9	5.5	6.5	3.5	7.4	7.0	7.1	11.7	13.2	12.6	21.8	16.2
Medical and dental specialists	6.0	c	10.1	5.9	15.0	11.4	8.7	c	15.3	15.5	c	18.2
Other technical and allied specialists	3.0	7.3	6.7	4.3	10.6	7.3	4.8	17.5	13.0	6.3	20.7	14.1
Administrative specialists and clerks	4.9	6.9	11.4	6.4	9.6	16.2	9.5	15.5	20.1	16.2	23.1	24.3
Electrical/mechanical equipment repairmen	4.0	5.2	7.0	4.3	6.6	10.3	5.4	10.8	13.0	9.9	16.6	13.0
Craftsmen	5.6	7.9	13.1	5.0	10.1	16.2	5.0	14.6	16.4	6.2	23.1	17.2
Service and supply handlers	15.3	12.7	19.3	15.2	17.7	20.9	10.4	23.3	20.7	14.3	30.4	23.8
Nonoccupational and miscellaneous[d]	5.8	9.0	10.9	12.6	24.4	10.9	13.4	16.5	12.0	15.7	18.6	13.5
Blacks as percent of all male enlisted personnel	5.9	8.7	10.0	6.4	13.7	12.6	8.0	17.0	14.6	11.7	21.9	16.2

Sources: Data for 1964 from Department of Defense, *The Negro in the Armed Forces: A Statistical Fact Book* (Office of the Deputy Assistant Secretary of Defense for Equal Opportunity, 1979). All other distributions derived from data provided by the Defense Manpower Data Center.
a. Percentage distributions are based on enlisted force composition as of December 1964, June 1972; and September 1976 and 1980.
b. Data for 1964 include both males and females.
c. The Navy provides the Marine Corps with medical support.
d. "Nonoccupational" includes patients, prisoners, officer candidates and students, persons serving in undesignated or special occupations, and persons who are not yet occupationally qualified (service members who are in basic or occupational training).

Appendix B   173

Table B-11. Blacks as a Percentage of Army Male Enlisted Personnel Assigned to the Twenty Most Common Occupational Subgroups, Fiscal Years 1972, 1976, and 1981[a]

Occupational subgroup in descending order of manpower strength[b]	1972	1976	1981
Infantry	21.4	24.5	29.2
Automative, general (repair)	12.3	18.0	30.4
Artillery and gunnery	17.8	31.3	45.0
Law enforcement (military police)	14.7	11.6	15.7
Supply administration	24.7	38.0	55.4
Armor and amphibious, general	11.8	16.1	25.9
Motor vehicle operator	19.3	21.7	35.6
Combat operations control, general	12.3	17.3	24.9
Medical care and treatment, general	19.8	24.2	34.7
Food service, general	28.6	33.2	46.0
Combat engineering, general	18.3	18.8	23.5
Unit supply	25.2	42.8	46.1
Missile artillery, operating crew	12.9	24.1	39.5
Linemen (wire communications)	26.3	39.3	57.8
Administration, general	14.9	33.0	40.9
Track vehicle repair	13.5	15.3	19.0
Personnel, general	13.6	27.9	44.2
Aircraft, general	8.6	11.1	16.0
Radio code (operator)	16.1	34.4	37.3
Communications radio (repair)	15.0	21.1	28.8
Blacks as percent of all male enlisted personnel	17.0	24.4	32.2

Source: Derived from data provided by Defense Manpower Data Center.

a. Percentage distributions are based on enlisted force composition as of June 1972 and September 1976 and 1981.

b. The order of occupational subgroups is based on the distribution of Army male enlisted personnel in September 1981. About 430,000 soldiers (approximately 70 percent of all male enlisted personnel in the Army) were assigned to these occupational subgroups at the end of fiscal 1981.

Table B-12. Black Male Enlisted Personnel in the Army, by Selected Primary Military Occupational Specialty and Occupational Area, September 1981[a]

Major occupational area and primary military occupational specialty code and title	Black male enlisted personnel	
	Number	Percent
Infantry and gun crews[b]		
11B/infantryman	15,459	29.2
11C/indirect fire infantryman	3,808	33.6
11H/heavy antiarmor weapons	1,979	24.5
13B/cannon crewman	10,881	46.9
19E/M48-60 A1/A3/armor crewman	4,727	26.1
19F/M48-M60 tank driver	346	24.7
12B/combat engineer	3,372	24.9
12C/bridge crewman	535	19.5
12F/engineer tracked vehicle crewman	332	25.0
16B/Hercules missile crew member	874	51.4
16D/Hawk missile crew member	1,165	47.6
16P/air defense artillery short-range missile crewman	1,184	44.1
16R/air defense artillery short-range gunnery crewman	1,105	35.0
16S/MANPADS crewman	1,193	41.0
Functional support and administration		
75B/personnel/administrative specialist	2,300	53.3
75C/personnel management specialist	805	54.3
75D/personnel records specialist	1,788	56.8
75Z/personnel senior sergeant	917	19.9
00E/recruiter	1,419	20.8
71L/administrative specialist	6,458	41.2
71D/legal clerk	346	25.0
71G/patient administrative specialist	507	51.5
00Z/command sergeant major	278	25.5
74D/computer machine operator	651	35.9
74F/programmer/analyst	175	15.8
73C/finance specialist	994	32.3
76C/equipment records and parts specialist	4,200	61.0
76J/medical supply specialist	584	53.5
76P/materiel control and accounting specialist	3,537	52.6
76V/materiel storage and handling specialist	2,618	59.9
76X/subsistence supply specialist	608	57.1
76Z/senior supply sergeant	331	27.1
76Y/unit supply specialist	8,360	46.2
71N/traffic management coordinator	678	41.2
71P/flight operations coordinator	325	32.5
71M/chapel activities coordinator	247	22.1
Service and supply handlers		
94B/food service specialist	8,789	46.8
64C/motor transportation operator	7,190	35.6
76W/petroleum supply specialist	2,370	63.8
95B/military police	3,420	15.8

Appendix B 175

Table B-12 (continued)

Major occupational area and primary military occupational specialty code and title	Black male enlisted personnel	
	Number	Percent
95C/correctional specialist	476	37.5
43E/parachute rigger	282	26.8
57E/laundry and bath specialist	480	64.1
57H/terminal operations coordinator	524	53.8
Communications and intelligence specialists		
05B/radio operator	1,208	28.3
05C/radio teletype operator	3,316	43.1
31M/multichannel communications equipment operator	3,631	52.8
17K/grand surveillance radar crewman	457	30.5
05H/EW/SIGINT Morse interceptor	247	18.6
98G/EW/SIGINT voice interceptor	61	3.5
98C/EW/SIGINT analyst	72	5.5
96B/intelligence analyst	243	16.2
13E/cannon fire direction specialist	774	24.1
13F/fire support specialist	1,147	23.3
19D/cavalry scout	2,875	26.0
31Z/communications-electronics operations chief	419	28.9
72E/telecommunications center operator	2,292	56.3
72G/data communication switching center specialist	1,144	44.6
Medical and dental specialists		
91B/medical specialist	5,378	36.1
91C/clinical specialist	1,499	31.1
91D/operating room specialist	334	30.1
92B/medical laboratory specialist	341	24.5
91E/dental specialist	523	39.3
91R/veterinary specialist	188	21.9
Other technical and allied specialists		
82C/field artillery surveyor	523	19.2
81C/cartographer	110	26.3
84B/still photographs specialist	196	28.7
55D/explosive ordnance disposal specialist	24	3.6
02B-02Z/musicians (all categories combined)	355	16.8
54E/chemical operations specialist	1,516	35.4
Electrical/mechanical equipment repairmen		
67G/airplane repairer	99	15.7
67N/utility helicopter repairer	744	15.5
67V/observation/scout helicopter repairer	448	18.0
67Y/attack helicopter repairer	354	15.6
68G/aircraft structural repairer	283	20.5
63B/light wheel vehicle/power generation mechanic	7,464	30.8
63W/wheel vehicle repairer	681	27.8
63H/track vehicle repairer	1,171	24.8

Table B-12 (continued)

Major occupational area and primary military occupational specialty code and title	Black male enlisted personnel	
	Number	Percent
63N/M60 A1/A3/tank system mechanic	624	17.6
63T/ITV/IFV/CFV/system mechanic	612	16.1
62B/construction equipment repairer	947	20.2
36C/wire systems installer/operator	3,418	59.1
36K/tactical wire operations specialist	5,785	57.6
45K/tank turret repairer	213	15.8
45N/M60 A1/A3/tank turret repairer	216	21.1
55B/ammunition specialist	1,108	48.5
52C/utilities equipment repairer	265	20.6
52D/power generation equipment repairer	481	21.6
Craftsmen		
44E/machinist	129	14.6
44B/metal worker	420	27.9
51B/carpentry and masonry specialist	604	25.7
62E/heavy construction equipment operator	404	15.4
62F/lifting and loading equipment operator	529	26.9
51N/water treatment and plumbing systems specialist	333	38.9
51R/electrician	200	22.8
51H/construction engineering supervisor	218	21.3
43M/fabric repair specialist	265	63.1

Source: Derived from data provided by Defense Manpower Data Center.

a. Approximately 87 percent of all black Army male enlisted personnel assigned to a primary military occupational specialty as of September 1981 are accounted for in this table.

b. Does not include service members assigned to special forces.

Table B-13. Sources of Commissions for Officers on Active Duty, September 1981

Source of entry[a]	White Number	Percent	Black Number	Percent	All officers[b] Number	Percent
Academy	32,141	12.2	841	5.5	33,567	11.8
ROTC						
Scholarship	28,183	10.7	1,346	8.8	29,949	10.5
Nonscholarship	50,181	19.1	3,798	24.8	54,822	19.2
OCS or OTS						
Direct procurement	18,140	6.9	635	4.1	19,028	6.7
In-service procurement	6,508	2.5	493	3.2	7,110	2.5
Procurement unknown	35,290	13.4	1,951	12.7	37,752	13.2
Direct appointment						
Physician or dentist	10,014	3.8	463	3.0	11,682	4.1
Other than physician or dentist	20,833	7.9	1,758	11.5	23,741	8.3
Aviation training program (not OCS or OTS)	14,957	5.7	268	1.7	15,560	5.4
Other or unknown	46,151	17.6	3,771	24.6	52,294	18.3
Total	262,398	100.0	15,324	100.0	285,505	100.0

Source: Derived from data provided by Defense Manpower Data Center. Percentages are rounded.
a. ROTC, OCS, and OTS are abbreviations for Reserve Officer Training Corps, Officer Candidate School, and Officer Training School, respectively.
b. There were 4,676 officers identified as members of other races. Also, the automated data files contained the records of 3,107 officers whose race could not be determined.

Table B-14. Reasons Given by Recruits for Joining the Military, by Race, 1979[a]
Percent of respondents

Reason	All reasons[b]		Most important reason[c]	
	White	Black	White	Black
Better myself in life	95.0	92.5	33.7	40.1
Get training for a civilian job	88.9	88.7	34.9	26.0
Serve my country	83.0	73.5	9.8	7.4
Travel and live in different places	77.3	74.7	4.5	4.1
Be away from home on my own	52.6	55.4	2.9	3.5
Get money for college education	50.1	53.7	7.2	9.9
Prove I can make it	47.1	49.3	2.9	3.8
Earn more money than as a civilian	30.6	34.8	1.2	1.3
Unemployed and could not find a job	13.5	20.7	1.6	3.0
Family tradition	9.4	11.1	0.3	0.4
Get away from a personal problem	6.5	8.0	1.0	0.5
Total	...	...	100.0	100.0

Source: Derived from data provided by Defense Manpower Data Center.

a. The 1979 Survey of Personnel Entering Military Service was administered to enlistees in all four military services, immediately following formal enlistment proceedings, at the Armed Forces Examining and Entrance Stations (AFEES). The survey was administered in two questionnaire variants and in two separate "phases." The first phase was conducted in March–April 1979; the second phase in September–October 1979. Data collection took place at all sixty-seven AFEES throughout the nation. The AFEES survey is conducted periodically for the purpose of gaining information on (1) the characteristics and experiences of men and women who enter military service and (2) the events and motivations that affect individual enlistment decisions.

b. The survey question was: "Below are some reasons that people join the military. Please tell us if each one is TRUE or NOT TRUE for you." The percentage frequencies shown here include all survey respondents who indicated that the stated "reason" was "true for me." The responses to separate questions—with identical wording but reversed ordering of response alternatives—on two different survey forms were combined to calculate the frequencies.

c. The survey question was: "Which of the reasons in Q.6 [previous question] is your *MOST IMPORTANT REASON* for enlisting? Mark One Letter Below." The responses to identical questions (but with ordering variations noted above) on two different survey forms were combined to calculate the percentage distributions.

Appendix B   179

Table B-15. **Percentage of Black and Other Minority Personnel in Selected Army Units, December 1980**

	Officers		Enlisted	
Unit designation	Black	Other minority	Black	Other minority
82nd Airborne Division	9.5	1.3	26.0	7.8
Battalions				
Infantry				
1	5.4	2.7	21.1	7.1
2	12.8	0.0	20.6	8.5
3	5.3	0.0	16.4	11.6
4	8.6	0.0	20.8	10.3
5	10.0	2.5	21.3	11.5
6	12.1	0.0	22.1	8.0
7	13.9	0.0	22.6	6.4
8	2.4	0.0	18.5	8.4
9	17.1	2.9	21.4	11.2
Armor	0.0	0.0	25.8	4.2
Artillery				
1	0.0	8.8	33.5	5.3
2	14.7	0.0	32.5	5.1
3	9.4	0.0	31.3	4.2
Air defense artillery	23.6	3.6	33.5	6.3
Air cavalry	9.2	0.0	28.6	5.3
Engineer	0.0	5.6	19.9	8.1
Signal	9.1	0.0	38.0	9.6
Medical	10.0	0.0	31.3	8.2
Aviation	6.8	0.0	24.8	6.6
101st Airborne Division (Air Assault)	9.8	1.0	33.8	4.8
Battalions				
Infantry				
1	12.1	3.0	29.4	6.0
2	3.1	3.1	36.2	3.5
3	9.7	0.0	33.8	5.3
4	18.8	0.0	31.9	4.6
5	9.1	3.0	37.6	3.9
6	16.1	0.0	32.1	5.1
7	6.3	3.1	29.9	6.7
8	12.5	0.0	34.3	5.5
9	6.1	0.0	32.1	4.4
Attack helicopter	3.6	0.0	30.5	6.8
Air assault	6.7	1.7	24.5	5.2
Artillery				
1	5.7	2.9	41.3	1.2
2	12.8	0.0	38.0	1.5
3	13.5	0.0	43.0	3.4
Air defense artillery	18.9	0.0	32.8	4.9

Table B-15 (continued)

Unit designation	Officers		Enlisted	
	Black	Other minority	Black	Other minority
Aviation				
1	5.9	2.9	31.7	4.7
2	9.7	0.0	35.0	4.7
Engineer	2.9	0.0	27.4	5.4
Signal	25.0	5.0	49.2	7.4
Medical	10.8	2.7	37.7	4.2
Transportation	0.0	0.0	27.6	5.8
3rd Armored Division	8.2	2.2	34.2	5.6
Battalions				
Infantry				
1	9.1	0.0	32.0	5.4
2	3.6	0.0	35.3	7.6
3	6.9	0.0	30.6	7.4
4	11.8	2.9	34.1	7.1
5	5.9	0.0	36.8	7.6
Armor				
1	3.1	9.4	29.6	4.7
2	9.4	3.1	35.0	6.6
3	10.3	0.0	30.4	3.5
4	10.3	0.0	28.8	8.5
5	9.1	3.0	28.8	4.7
6	7.4	0.0	29.1	5.2
Artillery				
1	2.7	0.0	41.8	4.7
2	9.5	0.0	40.2	7.8
3	12.9	3.2	42.1	3.9
4	6.1	3.0	40.2	4.1
Air defense artillery	16.3	5.9	38.5	3.3
Cavalry	3.3	3.3	29.1	7.7
Engineer	5.1	0.0	25.9	4.2
Signal	18.2	4.5	49.4	4.2
Medical	7.4	11.1	42.8	3.7
Aviation	3.8	1.9	30.6	4.9
2nd Infantry Division	15.6	2.3	41.1	6.3
Battalions				
Infantry				
1	27.8	0.0	45.9	5.9
2	23.1	5.1	37.9	6.1
3	15.4	2.6	42.9	9.1
4	17.5	2.5	41.7	6.1
5	15.4	2.6	37.9	7.9
Armor				
1	14.3	0.0	38.0	7.0
2	25.7	2.9	37.9	4.2
Artillery				
1	24.1	3.4	49.2	6.3

Table B-15 (continued)

	Officers		Enlisted	
Unit designation	Black	Other minority	Black	Other minority
2	16.1	0.0	46.6	3.5
3	27.6	0.0	50.7	5.8
4	2.9	0.0	43.7	4.6
Air defense artillery	21.1	0.0	40.9	3.9
Air cavalry	3.1	3.1	35.3	5.0
Engineers	8.8	2.2	31.9	10.3
Signal	9.5	4.8	57.6	5.0
Medical	23.8	7.1	46.5	5.5
Aviation	6.8	2.3	30.5	7.2
1st Cavalry Division	7.7	2.6	38.2	5.6
Battalions				
Infantry				
1	6.3	12.6	37.6	7.8
2	20.0	0.0	40.2	5.1
3	10.3	6.9	39.3	8.3
4	7.7	0.0	35.2	6.6
Armor				
1	5.6	2.8	37.4	4.0
2	3.6	3.6	37.4	5.1
3	9.1	3.1	33.2	5.4
4	5.9	2.9	33.3	5.8
Artillery				
1	17.2	0.0	48.5	5.0
2	11.1	2.8	44.2	5.5
3	5.7	0.0	43.2	5.2
Air defense artillery	25.0	0.0	43.1	2.9
Air cavalry	0.0	0.0	33.5	5.3
Engineer	3.6	0.0	29.1	6.7
Signal	4.3	4.3	53.6	3.8
Medical	4.1	0.0	46.9	7.2
Transportation	6.3	0.0	54.4	4.8
Aviation	3.7	3.7	30.7	5.1
197th Infantry Brigade	20.2	1.2	51.8	2.7
Infantry battalion 1	5.7	0.0	51.2	2.6
Company C	0.0	0.0	41.1	2.7
Infantry battalion 2	23.1	2.6	55.0	2.4
Company A	33.3	16.7	53.3	2.2
Armor battalion	16.1	0.0	48.3	2.2
Company B	25.0	0.0	50.0	2.9
Artillery battalion	18.2	0.0	59.9	2.7
C Battery	0.0	0.0	66.3	1.0
Combat support battalion	32.6	2.3	51.7	3.2
Engineer company	12.5	0.0	40.5	2.7

Source: Data provided by Department of the Army, May 1981.

Table B-16. Percentage of Black and Other Minority Enlisted Personnel in Marine Corps Units, December 1980

Unit designation	Black	Other minority
1st Marine Division	25.4	8.9
Battalions		
Infantry		
1	24.9	15.2
2	27.0	7.8
3	27.6	7.5
4	30.1	8.4
5	26.8	11.6
6	29.6	8.8
7	25.2	9.2
8	27.0	11.5
9	25.4	7.3
10	29.5	5.6
11	21.9	9.4
Artillery		
1	25.6	7.7
2	20.8	9.3
3	24.5	9.0
Tank	23.0	7.9
Assault amphibious	14.0	6.8
2nd Marine Division	33.0	2.4
Battalions		
Infantry		
1	32.5	2.3
2	37.6	2.4
3	37.1	2.6
4	33.6	2.6
5	32.9	3.3
6	38.2	3.1
7	37.2	2.9
8	35.9	2.8
9	34.8	2.3
Artillery		
1	35.7	1.7
2	36.2	2.8
3	34.4	1.5
4	32.6	1.3
5	36.4	1.7
Tank	27.1	3.1
Assault amphibious	12.8	2.0

Table B-16 (continued)

Unit designation	Black	Other minority
3rd Marine Division	29.6	4.8
Battalions		
Infantry		
1	28.1	4.2
2	32.1	3.8
3	29.7	4.2
Artillery		
1	31.7	5.7
2	35.2	6.7
Track vehicle	23.1	5.0

Source: Data provided by Headquarters, United States Marine Corps, May 7, 1981.

INDEX

Affirmative action: in Army occupation assignments, 55; backlash, 100; officer corps, 61
AFQT. *See* Armed Forces Qualification Test
Air Force: black representation, *1981,* 152; entry and occupational classification tests, 96, 97–98; exclusion of blacks, 18; integration, 30; racial incidents, Vietnam War, 36
Alexander, Clifford L., 6, 8n
Alexiev, Alex, 114n
Allen, William E., 92n
All-volunteer force: arguments for and against, 2–3, 9; attrition rate, 51–52; blacks in, 4, 6, 40, 152; entrance standards, 4; establishment, *1973,* 39; recruiting success, 10; representativeness, 7–9, 152, 153. *See also* Volunteers, black
Ambrose, Stephen E., 17
American Civil Liberties Union, 5n
American College Testing program, 122
Anderson, Bernard E., 68n, 70n, 101n
Anderson, Elijah, 101
Anderson, Martin, 82n
Angola, 114
Aptitude tests. *See* Scholastic Aptitude Test; Tests, standardized military
Arabs, 114
Armed forces: black representation, 41, 43, 152; employment opportunities for blacks, 33–34; and federal youth policies, 128–32; labor market effect on, 125–28; population representation in, 43–44; and population changes, 120–21; pressure for racial equality, 18–19, 26; proposed military buildup, 135; qualifying for, 121–25; racial imbalance, 38; racial integration, 11, 29–30. *See also* Air Force; All-volunteer force; Army; Marine Corps; Military effectiveness; Military manpower; Military service, blacks; Navy
Armed Forces Qualification Test (AFQT), 46–47, 87–90, 93–94
Army: black casualties, 15, 32, 76–79; black representation, 6, 14, 15, 41, 43–45, 152; blacks in Civil War, 13–15; blacks in World War I, 17–18; blacks in World War II, 20–21, 24–25; crime and prison statistics, 53–54; entry and occupational classification tests, 94, 97–98; integration, 29–30; need for skilled personnel, 133; occupational assignments to blacks, 55, 57; post-Civil War black regiments, 15; racial quotas, 7, 18, 27, 28; racial segregation, 18, 19, 27; recruitment goals, 4–5; resistance to integration, 27–28
AWOL, 52–53, 137
Azrael, Jeremy, 116n

Badillo, Gilbert, 1n
Barber, James A., Jr., 17n
Baskir, Lawrence M., 35n, 37n
Beard, Robin L., 106
Becker, Gary S., 70n
Berger, Jason, 9n
Binder, David, 6n
Binkin, Martin, 132n, 141n
Black Panthers, 112
Blau, Francine D., 69n
Blechman, Barry M., 134n
Board to Study the Utilization of Negro Manpower. *See* Chamberlin Board
Bock, R. Darrell, 93n, 94n, 96n
Bogart, Leo, 29n
Brehm, William K., 4, 5n
Breland, Hunter M., 122n
Brotzman, Donald G., 5
Brown, Charles W., 109n
Browning, Harry L., 72n, 74n

185

Brownsville race riot, 15–16
Brown, Terry W., 17n
Bureau of Colored Troops, 14
Butler, John Sibley, 36n, 41n, 84n, 107n

Callaway, Howard H., 4
Canby, Steven L., 63n, 80n
Carmichael, Stokely, 76
Carney, Larry, 44n, 137n
Carpenter, M. Kathleen, 100n, 141n
Carper, Jean, 33n
Casualties, black: Civil War, 15; concern about, 80–83; predicted for future wars, 78–80; Vietnam War, 32, 76–78
Chamberlin Board, 27–28n
Chisholm, Shirley, 81
Civilian Advisory Panel on Military Manpower Procurement, 33n
Civil War: blacks in military service, 13–14; draft, 64n; mortality rate of black soldiers, 15
Clark, Blair, 2n
Cockerham, William C., 113n
Coffey, Kenneth J., 4n, 5n, 41n, 44n, 100n, 112n, 115n
Cohen, Lawrence E., 113n
Colonial America, blacks in militia, 11–12
Committee for the Study of National Service, 150
Condran, John G., 100n
Congressional Budget Office, 127
Congressional Research Service, 40
Congress of Racial Equality (CORE), 76–77
Conscription. *See* Draft
Conscription Act, *1863*, 14
Cooper, Richard V. L., 9n, 44n, 125n
Cornish, Dudley Taylor, 12n, 15n
Court-martials, 54
Crane, Elaine, 76n
Crime rates, Army, 53
Cuba, 110–11, 114
Curry, G. David, 1n

Dalfiume, Richard M., 18n, 19n, 25n, 28n, 30n
David, Jay, 76n
Davis, James W., Jr., 2n, 35
Dayan, Moshe, 118n
Defense, Department of, 6n, 24n, 46n, 47n, 51n, 67n, 73n, 93n; and all-volunteer force, 9; enlistment standards, 4; handling of off-base discrimination, 31; study of manpower requirements, 40
Defense Manpower Commission, 40, 99
Dellums, Ronald V., 5–6n, 82

Denton, Herbert H., 61n
Department of Defense Authorization Act, *1981*, 137
Desertion, 52–53
Detroit civil disorder, *1967*, 109
Discharge, black versus white rates of, 55
Doering, Zahava D., 104n
Dolbeare, Kenneth M., 2n, 35
Dominguez, Jorge I., 114n
Dominican Republic, 115, 118
Draft: arguments against, 39–40; arguments for, 2, 64, 65; avoidance of, 33; blacks in Vietnam War, 32–33; Civil War, 14; economic discrimination in, 35–36; individualism and, 64–65; for national service, 149–51, 157; potential racial mix under selective, 145, 147–48; racial inequity in, 1, 2, 75; revival of, 8, 9–10, 145, 147. *See also* Selective Service System
Dred Scott decision, 26n
Drummond, Bill, 54n
Dubois, W. E. B., 17

Earnings in military, 71
Educational level: black versus white recruits, 47–48; as predictor of military effectiveness, 86–87
Education, Department of, 125n
Edwards, Don, 6n
Eisenman, Richard L., 144n
Eitelberg, Mark J., 25n, 41n, 44n, 92n
El Salvador, 114–15n
Emancipation Proclamation, 13
Employment–population ratio, 69–70
Equity, racial: all-volunteer force and, 2; draft and, 1, 2, 75; military service and, 62–63, 83
Everett, Robinson O., Jr., 54n

Fahy, Charles H., 26
Fahy Committee, 26–28, 43
Fallows, James, 107
Fauntroy, Walter E., 82
Federal Bureau of Prisons, 54
Fletcher, Marvin, 15n, 16n, 23n
Foner, Jack D., 12n, 14n, 15n, 17n, 19n, 24n, 30n, 110n, 111n
Franklin, John Hope, 15n
Freeman, Richard B., 70n
Friedman, Milton, 63
Fullinwider, Robert K., 83n

Gates Commission, 2–3, 4, 9n, 39, 40, 43, 80–81
Gates, Thomas S., Jr., 2
Gatewood, Willard B., Jr., 111n

Index **187**

General Educational Development program, 67
Gerhardt, James M., 33n, 75n
Gesell Committee, 31–32
Gesell, Gerhard A., 31
GI Bill, 142, 144
Gibson, Truman, 22
Gillem, Alvan, 22
Gillem Board, 22, 27
Ginsburgh, Robert N., 116n
Ginzberg, Eli, 23
Goldberg, Lawrence, 125n
Goldich, Robert L., 40n
Good, Robert C., 115n
Grant, James, 129n
Gray, Gordon, 28
Greenberg, I. M., 91n
Green, Robert L., 92n
Griffore, Robert J., 92n
Grimshaw, Allen D., 16n
Grissmer, David W., 125n, 126n, 139n
Gropman, Alan L., 27n, 31n, 36n
Grundy, Kenneth M., 113n
Guinn, Nancy, 92n

Haig, Alexander M., Jr., 114n
Hamermesh, Daniel, 129n
Harden, Blaine, 107n
Harding, Warren G., 17n
Haveman, Robert H., 132n
Hershey, Lewis, 1
Hesseltine, William B., 14n
Hicken, Victor, 26n
Hoffman, Martin R., 106
Hoiberg, Anne, 86n
Hope, Richard O., 22n
Houston race riot, 16
Houstoun, Marian F., 129n
Huck, Daniel, 139n
Hutzler, William P., 104n

Indian Wars, 15
Integration. *See* Racial integration of armed forces
Iran, 118

Janowitz, Morris, 26n, 41n, 82n, 102n, 106n
Johnson, George E., 130n
Johnson, Thomas A., 77n
Jordan, Vernon, 82
Jurkowitz, Eugene L., 74n

Karpinos, Bernard D., 137n
Keeley, John B., 107n
Kelley, Joseph, 9n, 46n
Kennedy, Edward M., 2n, 81

Kerner Commission, 109
Kim, Choongsoo, 51n
King, Martin Luther, Jr., 76
Kohen, Andrew I., 51n
Korb, Lawrence J., 8n
Korean War, 28; Chinese appeal to black troops, 117–18; and integration, 29, 30, 102
Ku Klux Klan, 107
Kyriakopoulos, Irene, 132n, 141n

Labor market for young blacks: federal social programs, 129–32; segmentation, 128; substitution, 128–29; two-tier minimum wage, 130–31; wage subsidies, 131–32
Laird, Melvin R., 1n, 3
Larson, Gerald E., 122n
Latin America, 114–15
Lazarsfeld, Paul F., 101–02n
Lee, Ulysses, 14n, 15n, 19n, 21n, 22n, 23n, 24–25, 30n, 77n
Leinwand, Gerald, 1n, 39n
Lerman, Robert I., 132n
Lincoln, Abraham, 14
Lister, Sara E., 90n
Little, Roger W., 14n
Lopreato, Sally C., 72n, 73n
Lord, Sharon, 100n

McCloskey, Paul N., Jr., 81–82n, 151
MacGregor, Morris J., 22n, 23n, 28n
Malcolm X, 112
Mandelbaum, David G., 101n, 102n, 103n
Mare, Robert D., 128n
Marine Corps: black representation, 6, 152; blacks in World War I, 18; blacks in World War II, 20, 24; integration, 30; need for skilled personnel, 133; potential black casualties in future wars, 79–80; racial incidents, Vietnam War, 36; racial quotas, 5
Marmion, Harry A., 2n, 14n, 35n
Marshall Commission, 33, 77
Marsh, John D., Jr., 6–7
Martin, Harold H., 29n
Mason, William, 74n
Mathofer, Hans, 118n
Meckler, Alan M., 63n
Meese, Edwin, III, 114n
Merton, Robert K., 101n
Middleton, Lorenzo, 108n
Military effectiveness: determinants of, 84; domestic racial allegiances and, 108–10, 156; educational level and, 86–87; entry test standards and, 89–98; foreign percep-

188 Blacks and the Military

tions of, 116–19, 156; indicators of individual, 85–86; international racial allegiances and, 110–15, 156; minority group relationships and, 98, 155; racial integration and, 102–03; racial segregation and, 103; socioeconomic status and, 99–100
Military manpower: educational benefits, 141–44; enlistment standards, 85, 136–38, 154–55; pay increases, 138–39, 141; and technological developments, 132–35. *See also* Occupations, military
Military Manpower Task Force, *1981*, 158
Military service, blacks, 11; benefits, 63, 65–75, 154; as bridge to better life, 72–75, 154; casualty risks, 75–83, 154; in Civil War, 13–17; debate over recruitment, 63; earnings, 71; as education alternative, 66–67; equity arguments, 62–63, 83; in Korean War, 28–30; peacetime versus wartime, 62; in Revolutionary War, 12–13; as source of employment, 65–66, 153; in Vietnam War, 32–37; in World War I, 17–18; in World War II, 19–25. *See also* All-volunteer force; Draft; Volunteers, black
Miller, James C., III, 63n
Millis, Walter, 64n
Milton, H. S., 24n, 25n, 27n, 29n, 30n, 57n, 102n
Minckler, Rex D., 116n
Minimum wage, two-tier, 130–31
Mislevy, Robert J., 93n
Moore, Elsie G. J., 94n, 96n
Moore, Harold G., 5n
Moskos, Charles C., Jr., 29n, 30n, 31n, 35, 36n, 41n, 49n, 82n, 83, 102n, 106n, 107n, 108n, 109n, 115n, 118n, 141n
Moynihan, Daniel P., 33–34, 112n
Mullen, Robert W., 36n, 110n, 111n, 112n, 117n
Murray, Paul T., 77n
Myrdal, Gunnar, 16n, 18n, 65, 66n

Nalty, Bernard C., 28n
National Advisory Commission on Selective Service. *See* Marshall Commission
National Assessment of Educational Progress, 122–23, 125n
National Guard: racial policy, 32; riot duty, 109
National security: ambivalence about military service and, 157–58; armed forces racial composition and, 84; military recruitment and, 2, 39, 63; representativeness of armed forces and, 153; social services versus, 152, 159. *See also* Military effectiveness
National service, 149–51, 157–58
National Service Act, proposed, 151
National Urban League, 77
Navy: black representation, 14, 18, 152; blacks in Continental, 13; blacks in World War I, 18; blacks in World War II, 20, 24; integration, 30n; need for skilled personnel, 134; racial incidents, Vietnam War, 36; racial quotas, 5
New York City race riot, 14
Nichols, Lee, 17n, 21n, 22n, 29n
Nigeria, 119
Niles, David K., 28n
Nixon, Richard M., 1, 39
Nordlie, Peter G., 87n, 89n
North, David S., 129n
Northrup, Herbert R., 55n
Novak, Michael, 25n

Occupations, military: affirmative action goals in assignments, 55, 57; aptitude tests, 55, 94; blacks in specialty, 57–58; factors influencing performance, 84–85; promotions, 58–59; technical skill requirements, 132–35; transferability to civilian life, 74–75
Officer corps, black underrepresentation in, 59, 61, 152, 153
O'Mara, Francis E., 108n
Osterman, Paul, 128n
O'Sullivan, John, 63n

Palmer, John L., 131n, 132n
Parnes, Herbert S., 51n
Pauly, Mark V., 63n
Pear, Robert, 129n
Pechman, Joseph A., 126n
Pell, Claiborne, 81
Pershing, John, 15
Phelps, Edmund S., 71n
Philippines, 111–12
Popham, W. James, 90n
Population changes, 120–21
Poston, Dudley L., Jr., 72n, 73n
Post-Vietnam Era Veterans' Educational Assistance Program (VEAP), 142
President's Commission on an All-Volunteer Armed Force. *See* Gates Commission
President's Committee on Equality of Treatment and Opportunity. *See* Fahy Committee
President's Committee on Equal Opportunity in the Armed Forces. *See* Gesell Committee
Primary occupational specialty, 58–59
Prisons, 54–55

Index 189

Project Clear, 29–30
Project One Hundred Thousand, 34, 90, 91
Public opinion polls: on military draft, 145n; on racial tension, 100
Purnell, Karl H., 32n

Quarles, Benjamin, 12n, 13n, 14n
Quotas, military racial, 2, 5, 7, 20, 27

Race relations: soldiers' attitudes, 103–05; "tipping point" theory, 105–08
Racial consciousness of blacks, 37, 101
Racial integration, armed forces, 11; Korean War test of, 29–30; and military effectiveness, 102–08
Racial segregation, 18; in Army, 19; black pressure to eliminate, 25; and military effectiveness, 103; World War II mobilization and, 20
Racism: as backlash to affirmative action, 100; in communities near military bases, 31–32; institutional, 32, 33
Reading ability assessments, 122–23
Reagan administration: and draft, 145; economic program, 126; projected unemployment rate, *1980s*, 127; proposed cutbacks in youth programs, 129–30, 144n; proposed military buildup, 135
Reagan, Ronald, 9, 158
Rebh, Richard G., 116n
Reeg, Frederick J., 5n, 41n, 44n
Reserves: black representation, 43; racial policy, 32
Revolutionary War, blacks in militia, 12–13, 64
Rimland, Bernard, 122n
Riots, race: Brownsville, 15–16; Houston, 16; and National Guard, 109; New York City, 14; and regular Army, 110
Robey, Bryant, 121n
Roosevelt, Franklin D., 19
Rowan, Carl T., 118n
Russell, Richard B., 26–27n

Sarkesian, Sam C., 86n
Sawhill, Isabel V., 68n, 70n
Schexnider, Alvin J., 41n, 84n
Scholastic Aptitude Test (SAT), 122, 123
Segregation. *See* Racial segregation
Selective Service System, 1, 147n; and racial identification, 20; treatment of blacks, Vietnam War, 33–35, 77
Selective Training and Service Act of *1940*, 18–19
Shields, Mark, 8n
Shils, Edward A., 101n
Shuler, Edgar A., 16n

Sims, William H., 87n
Skill qualification test (SQT), 58–59, 85
Slaves: in colonial army, 11–12; as substitute draftees, 13
Smith, Marvin M., 71n
Socioeconomic status: black volunteers, 49, 51; military effectiveness and, 99–100
South Africa, 113
Soviet Union, 114n; minorities in military, 116–17, 156
Special Board on Negro Manpower. *See* Gillem Board
Stern, Sol, 76n
Stillman, Richard J., II, 18n, 30n, 32n
Stone, Marvin, 10n
Stouffer, Samuel A., 22n, 30n, 101n
Strauss, William A., 35n, 37n
Student aid programs, 143
Student Non-Violent Coordinating Committee (SNCC), 77
Survey of Personnel Entering Military Service, *1979*, 67
Sussman, Barry, 61n

Targeted Jobs Tax Credit program, 131–32
Taylor, Maxwell D., 8n, 117n
Tella, Alfred, 69n
Tests, standardized military, 67; errors of rejection, 90–91; and military performance, 89–90; racial bias, 92–93; racial comparison of scores, 46–47, 93–98; subtests for specific occupations, 94; and trainability, 89–90, 91
"Tipping point" theory, 105–08
Tollison, Robert D., 35n, 63n, 77n
Truman, Harry S., 18, 26, 28
Truss, Ann R., 87n
Tupes, Ernest C., 92n

Unemployment, youth, 67; of blacks versus whites, 68–69; cutbacks in employment and training programs, 129–30; and enlistments, 125–26; projected rate, *1980s*, 127; race discrimination, 70–71; two-tier minimum wage, 130–31; understatement of black, 69; wage subsidies, 131–32

Valentine, Lonnie D., 92n
Veterans Administration, 142
Vietnam War: black casualties, 32, 76–78; draft of blacks, 32–36; racial incidents, 36–37
Volunteers, black: concentration in lower ranks, 58; disciplinary incidents, 52–53, 59; educational level, 47, 49, 59; enlistment standards, 47n, 55, 137–38; performance on military aptitude tests, 46–

47; reasons for enlistment, 67; socioeconomic status, 49, 51

Wachter, Michael L., 68n, 70n
Wage subsidies, 131–32
War Department, 19, 21n
Washington, Booker T., 111
Washington, George, 12, 64
Waters, Brian K., 122n
Wedekind, Lothar H., 107n
Weigley, Russell F., 63n, 64n, 65n
Weinberger, Caspar, 135
Weinraub, Bernard, 9n
Weinstein, Paul A., 74n
Westcott, Diane M., 128n
West Germany, 118
Wexler, Jacqueline Grennan, 149n
Wharton, Yvonne L., 122n
Wilkins, Roger, 101
Wilkins, Roy, 112
Willett, Thomas D., 63n
Williams, Eddie, 99n
Wilson, George C., 5n, 6n
Wilson, Kenneth L., 36n, 107n
Wilson, Margaret Bush, 113
Wilson, Woodrow, 17n
Wimbush, S. Enders, 114n
Winship, Christopher, 128n
Wirtz, Willard, 122n
Wofford, Harris, 149n
Women in military: disciplinary incidents, 53; separation rate, 55
Work incentive (WIN) tax credit program, 131–32
World War I: blacks serving in, 17–18; draft, 65; German appeal to black troops, 117
World War II: appeal for black combat volunteers, 20–21; draft, 65n; Japanese appeal to black troops, 117; military's racial policy in, 19–20; military's postwar racial policy, 25–28; number of blacks serving in, 24; performance of black soldiers, 21–24
Wright, Walter L., Jr., 23–24

Yarmolinsky, Adam, 36n, 109n
Young, Whitney M., Jr., 35, 112
Youth Education and Demonstration Projects Act, 129